Lubricating Polymer Surfaces

T0321098

YOSHITO IKADA, Ph.D.
YOSHIKIMI UYAMA, Ph.D.

Research Center for Biomedical Engineering
Kyoto University

CRC Press
Taylor & Francis Group
Boca Raton London New York

CRC Press is an imprint of the
Taylor & Francis Group, an **informa** business

Lubricating Polymer Surfaces

CRC Press
Taylor & Francis Group
6000 Broken Sound Parkway NW, Suite 300
Boca Raton, FL 33487-2742

First issued in paperback 2019

ISBN-13: 978-1-56676-013-3 (hbk)
ISBN-13: 978-0-367-40046-0 (pbk)

Visit the Taylor & Francis Web site at
http://www.taylorandfrancis.com

and the CRC Press Web site at
http://www.crcpress.com

Main entry under title:
  Lubricating Polymer Surfaces

A Technomic Publishing Company book
Bibliography:
Includes index p. 167

Library of Congress Catalog Card No. 93-60803

Preface   *vii*

Introduction   *xi*

Chapter 1. Principles of Friction and Lubrication 1

1.1 Friction   1
1.2 Lubrication   8
1.3 Wear   13
1.4 References   15

Chapter 2. Test Methods for Lubricated
Polymer Surfaces 17

2.1 Methods for Determining Coefficients of Friction   17
2.2 Methods for Determining Wear   17
2.3 Methods Used in Practice   22
2.4 Medical Devices   28
2.5 References   39

Chapter 3. Lubricious Polymer Surfaces 41

3.1 Characteristics of Polymer Surfaces   41
3.2 Frictional Properties of Polymers   42
3.3 Lubrication and Slipperiness   45
3.4 Slippery Polymer Surfaces   45
3.5 References   53

Chapter 4. Lubricated Surfaces for Medical Use 55

4.1 Guidewire   56
4.2 Catheter   58
4.3 Artificial Joint   61
4.4 Suture   63
4.5 Contact Lens   65

4.6 Orthopedic Casting Tape   66
4.7 Surgical Glove Powder   67
4.8 Vial Stopper and Others   68
4.9 References   69

Chapter 5. General Methods for Surface Modification **73**

5.1 Polymer Surface Structure   73
5.2 The Purpose of Polymer Surface Modifications   75
5.3 Chemical Modification   78
5.4 Physical Modifications   79
5.5 Biological Modifications   88
5.6 References   88

Chapter 6. Surface Analysis of Modified Polymers **91**

6.1 Instability of Polymer Surfaces   91
6.2 Methods for Analyzing Functional Groups on
    Polymer Surfaces   92
6.3 Contact Angle Methods   95
6.4 Spectroscopic Methods   102
6.5 ATR FT-IR   107
6.6 References   109

Chapter 7. Surface Grafting **111**

7.1 Grafting by Coupling Reaction   112
7.2 Graft Polymerization   113
7.3 References   133

**Chapter 8. Surface Structures and Properties of
Grafted Polymers** **139**

8.1 Structures   139
8.2 Surface Properties   143
8.3 References   162

Appendix   *165*

Index   *167*

Large-volume polymers are becoming increasingly important in the world economy. Let us take a look, for example, at polyethylene terephthalate (PET). Recently, PET resins have been used increasingly in plastic containers, cases, and bottles, and PET film has been widely applied in many fields, such as in electrical engineering for magnetic recording tapes, magnetic memory cards, other computer-assisting tapes, electrical motor slot liners, insulators for wire and cable, transformers, capacitors, and so on. In other fields, book jackets, stationery supplies, pipe wrap, and photographic film bases are also usually made of PET. Although PET has proven to have the most excellent mechanical properties among the commodity polymeric materials, problems have often been pointed out with respect to its surface properties. In almost all cases, it is necessary to modify the surfaces of PET materials to give them properties such as printability, paintability, wettability, antifogging, antistatic, antistaining, antifouling, and lubrication equal to those of other polymers. When polymers are used as textiles, it may also be important to increase wrinkle and flame resistance, wettability, and dyeability, in addition to the properties listed above.

Therefore, a variety of surface modification techniques have been proposed not only for PET but also for other synthetic and natural polymeric materials. Although many papers have been published describing the methods for improving the poor surface properties of commodity polymers, there have been relatively few monographs dealing with the technologies for surface modification. Examples include *Modification of Polymers*, edited by E. Charles, J. Carraher, and M. Tsuda (ACS Symposium Series Vol. 121, 1980), and *Surface Preparation Techniques for Adhesive Bonding*, by R. F. Wegman (Noyes, 1989).

On the other hand, numerous textbooks and studies have been published on the surface analysis or surface chemistry of polymeric materials. Some of these include *Physical Chemistry of Surfaces*, by A. W. Adamson (Wiley, 1976); *Polymer Surfaces*, edited by D. T. Clark and W. J. Feast (1978); *Surface and Colloid Science*, edited by R. J. Good and R. R. Stromberg (Plenum, 1979); *Polymer Surfaces*, by B. W. Cherry (Cambridge University Press, 1981), *Colloid and Interface Chemistry*, by R. D. Vold and M. J. Vold (Addison Wesley, 1983); *Physicochemical Aspect of Polymer Surfaces*, edited by K. L. Mittal (Plenum, 1983); *Surface and Interfacial Aspects of Biomedical Polymers*, edited by J. D. Andrade (Plenum, 1985); *Polymer Surfaces and Interfaces*, edited by W. J. Feast and H. S. Munro (Wiley, 1987); *Polymer Surface Dynamics*, edited by J. D. Andrade (Plenum, 1988); and *Polymer Surface and Interfaces II*, edited by W. J. Feast, H. S. Munro, and R. W. Richards (Wiley, 1993).

Indeed, the surface modification of polymeric materials has been the object of a large number of investigations, but little attention has been paid to making a polymer surface frictionless or slippery, and lubricating surfaces are practically unmentioned in any books so far published, probably because of the relatively minor importance of polymer friction in industrial applications. Lubrication has been widely considered to be important in metals and ceramics, but not with polymers, either in application or in theory. Although lubrication always accompanies wear—against which polymers are quite weak, especially under large normal loadings—a lubricating polymer surface is important, especially in marine and biomedical technologies. For instance, biomaterials to be used for catheterization on the urinary, tracheal, and cardiovascular tracts, or for endoscopy, should have a surface with good handling characteristics when dry and which preferably becomes slippery upon contact with body liquids. Such a low-friction surface must enable easy insertion and removal of the device from a patient. It would further prevent mechanical injury to the mucous membranes and minimize discomfort to the patient. Earlier approaches to providing a low-friction surface were mostly simple applications involving lubricants such as lidocaine jelly, silicone oil, or non-permanent coating with low-friction materials such as polyethylene or fluoroplastics. However, these substances cannot maintain a high degree of slipperiness for the required duration of time, due to the fact that they leach or disperse into the surrounding body fluid.

Although very few man-made materials are known that have slippery surfaces, it is not difficult to find such surfaces in the biological world. For instance, most of the surfaces of fish and sea wood are very slippery when brought into contact with water. There are also many slippery surfaces on mucous tissues and organs in animal bodies. The purpose of this slipperiness in nature has not yet been fully understood, but it is obvious that the slippery surface greatly reduces the frictional resistance occurring when the surface slides on another solid object. In most cases, these slippery surfaces have a large content of water, which acts as a lubricant. The surface may feel slippery when water is squeezed out by a fingertip touch. Another interesting feature of such slippery surfaces is to prevent the surface from adhering to other objects with which it is in contact. All of the naturally occurring slippery surfaces are extraordinarily hydrophilic, but polymers such as cellulose, poly(2-hydroxyethyl methacrylate), and poly(vinyl alcohol), do not become nearly as slippery as the natural organisms do when they are brought into contact with water, although these polymers are known as hydrophilic.

The aim of this book is to describe the principle of lubrication, to outline a variety of methods for attaining a lubricious surface, and to describe the characteristics and properties of such lubricious surfaces. The technology for surface modification of polymers by grafting will find other applications than for lubrication, such as for improvement of the interfacial adhesion in polymer composites. We are grateful for many fruitful discussions and valuable advice from Dr. Daniel Graiver, Dow Corning Corp., Prof. Ken Ikeuchi, Prof. Masanori Oka, Mr. Koichi Kato of Research Center for Biomedical Engineering, Kyoto University, Dr. Kang En-Tang, University of Singapore, and Dr. Ko-Shao Chen, Department of Materials Engineering, Tatung Institute of Technology.

Much attention has been directed to the practice and theory of the lubrication of metals and ceramics. The same cannot be said for the lubrication of polymer surfaces, a topic which seems to have gone unnoticed for a long time in the literature of materials engineering. The reasons, we believe, are twofold: the use of polymers is less prevalent than the use of metals and ceramics; the lubrication of polymers is more complex than that of metals and ceramics.

It is more difficult, for example, to measure the frictional properties of polymers because friction is greatly influenced both by the part in direct contact with the polymer, and by the environmental medium.

Theoretical and practical problems associated with the lubrication of polymer surfaces are concisely described in Chapter 1, as are the general principles of lubrication and friction for nonpolymeric, hard materials.

Chapter 2 summarizes the practical methods currently used to determine the coefficient of friction, including the standard ASTM method. It is normal practice to determine the coefficient of friction of materials in order to characterize their surface lubrication. This becomes difficult with the complicated shapes of many molded products.

Lubricious polymer surfaces are especially important in the field of biomedical engineering. Biomaterials used for catheterization and endoscopy, for example, should quickly become slippery upon contact with bodily fluids. This allows easy insertion or removal of medical devices into and out of patients. Earlier approaches to lubrication were to simply apply lubricants such as silicone oil to the surface of the device or to coat it with low-friction materials such as polyethylene and fluoroplastics.

As noted in Chapter 2, a surface has a very low coefficient of friction if its surface energy is either extremely high or low, i.e., if the water contact angle is either very high or low. From a practical point of view, it is easier to achieve an extremely hydrophilic polymer surface than it is to achieve an extremely hydrophobic one, because the water contact angle of even the most hydrophobic surface currently available is no larger than about 120. Thus, hydrophilic coatings seem to be a very popular choice for lubricating polymer substrates, with a dip-coating process the simplest and most widely employed.

Because the lubrication of coated surfaces is greatly influenced by the water content of the coated material, many efforts have been made to prepare a surface or a material which can absorb large amounts of water molecules without altering the substrate structure. Various methods for producing these hydrophilic lubricious coatings are summarized in Chapter 3.

Lubricious polymer surfaces find wide applications in any field related to aqueous liquids. These include medical, biotechnical and marine science and technologies. In Chapter 4 we describe published literature, including patents, on medical devices associated with lubricious surfaces. These devices include guidewires, catheters, artificial joints, sutures, contact lenses, orthopedic casting tapes, surgical gloves, and so on.

To develop better lubricated polymer surfaces than those currently available, it is essential to understand at least the phenomena related to polymer surfaces, and to have a good knowledge of current surface modification techniques. Thus various methods for surface modification — not necessarily limited to lubricated polymer surfaces — are reviewed in Chapter 5. The surface modification methods are classified into two major groups — chemical modification (wet process), and physical modification (dry process), such as plasma treatment and irradiation with ionizing and nonionizing radiation.

Although a variety of analytical methods are currently available for characterizing polymer surfaces, there are some problems that are mostly associated with the instability of polymer surface structure. The polymer surface is susceptible not only to oxidation and hydrolysis, but also to structural change due to micro-Brownian motion of the surface segments. It is always important in surface analysis to know to what depth the analytical means is intended to probe. The contact angle method is still popular, because this simple measurement gives valuable information on the properties of the outermost thin layer of polymer. This information includes hydrophilicity-hydrophobicity balance, and the overturn of polar caps at the surface. In Chapter 6 we deal with methods of surface analysis which are relatively new, and often used for modified polymer surfaces.

Where a surface is to be used in contact with water, surface grafting of water-soluble polymers is more effective than hydrophilic coating as a method for obtaining a sufficiently lubricious surface. However, little attention has been directed toward surface grafting, even though this method seems to be very promising for modification of polymer surfaces and has a number of possible applications. The limited number of studies currently conducted in this field is probably due to the difficulty of the reactions and the assessment of the products. Representative initiation methods for surface graft polymerization are described in Chapter 7.

A polymer surface modified by the graft polymerization of water-soluble monomers possesses unique properties in aqueous environments. One of the most outstanding properties is a high level of lubrication. On the other hand, the polymer surface with grafted water-soluble chains undergoes substantial adhesion to another substrate surface when they are brought into contact in the presence of water under pressure and then dried.

Even in a fully hydrated environment, grafted water-soluble polymer chains gradually lose water molecules from the region of contact, resulting in an increase both in the frictional force caused by the increased viscosity, and in the adhesive force. As the graft density increases, abundant molecules can be absorbed, and they play an important role as a fluid-film lubricant.

Finally, it should be noted that drag-reduction systems probably apply to lubricated polymer surfaces. Although a few efforts have been successful in improving turbulent flow, there are still unresolved problems in drag-reduction systems.

CHAPTER 1

Principles of Friction and Lubrication

Lubrication, as well as wear, has been of great technological importance, especially in the field of machinery, and most of the investigations of lubrication have focused on lubricants applied to bearings and gears made of steel. Less attention has been directed to lubrication phenomena in the field of polymer science, since polymeric materials have been thought to be soft and poor in resistance against wear. Indeed, polymers have been used only as additives, such as solid-film lubricants. However, recent progress in polymer technology has enabled partial substitution of polymers for metallic materials, resulting in concurrent changes in lubrication science. Since detailed explanations for lubrication phenomena can be found in many monographs [1−6], only a brief description will be given in this chapter.

1.1 FRICTION

1.1.1 Hard Materials

When a solid material is allowed to slide over another surface, friction drags are inevitably caused, both in the initial stage of motion (static friction) and during continuous traveling (kinetic friction). The friction generally results from the force or energy required to shear the neighboring substance and deform contacting substances between the two surfaces, regardless of their phase, i.e., vapor, liquid, or solid. Therefore, the frictional force is closely related to the mechanical properties, especially the shear strength of the material in contact. Usually, the friction of solid materials is much larger than that of liquids and gaseous materials, and so lubricants are commonly used in viscous liquid or oil form. Theoretically, the lowest friction of all will be achieved when two opposing solid materials are separated by an air jet, as depicted in Figure 1.1.

The term "friction" should be distinguished from the term "lubrication" to avoid confusion in understanding the lubrication phenomena. As described above, friction is inevitably caused by shearing two contacting

1

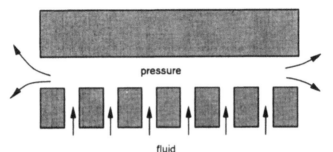

FIGURE 1.1. A simple model of an externally pressurized hydrostatic or aerostatic lubrication.

materials, irrespective of their phase, whereas lubrication is a way to lessen the frictional force between the two materials, for instance, by utilizing lubricants.

To assess the effects of lubricants or to determine the degree of slipperiness or lubrication between two solid materials, a coefficient of friction (μ) is widely used. This coefficient is a nondimensional parameter, defined by the ratio of the frictional force to the weight or the load applied onto the material. The law of friction was first stated by Leonardo da Vinci, who showed that the force needed to initiate or maintain sliding motion is proportional to the normal loading. Amontons and Coulomb also attempted to explain the frictional behavior, and concluded that the force of friction should be attributed to the interlocking of local surface roughness and asperities of the two solid surfaces [7,8]. According to Coulomb, the μ value is defined as:

$$\mu = F/W \qquad (1.1)$$

where F and W are the horizontal force component and the normal load applied, respectively (the second law of friction). The first law of friction states that μ does not depend on the apparent area of contact. Coulomb introduced further a third law, which states that friction is independent of the sliding speed. This implies that the force required to maintain motion at any specified velocity is constant. In many practical situations this is not generally true, although the second law has been confirmed in most experimental results. As an example, the μ value observed for a steel-aluminum countersurface is plotted in Figure 1.2 against the load applied on the system [9]. As can be seen, μ is constant over a wide range of loads from 10 mg to 10 kg. Table 1.1 summarizes the μ values determined for different pairs of metallic materials [10].

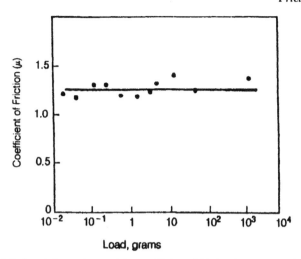

FIGURE 1.2. Load dependence of the coefficient of friction between steel and aluminum countersurface. (Source: Whitehead, 1950 [9].)

A brief attempt was made to predict a μ value between two metals in their dry state based on their mechanical properties [1]. Briefly, when a simple system consisting of a hard material (steel) and a soft metal (copper) is subjected to compression (Figure 1.3), the softer material (copper) acts as a lubricant. In contrast, no such lubrication occurs in a system where the contacting surfaces are both hard, such as in a steel/steel contacting system. When the vertical force of the compression increases, the steel will undergo elastic deformation, followed by plastic deformation which remains permanently. The shape of the steel changes outwards due to the plastic

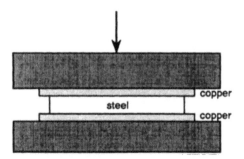

FIGURE 1.3. A simple model of lubrication by a soft solid. A steel block flows outwards as it is compressed, and the relative sliding is facilitated by shear within the soft copper. (Reproduced from *Lubrication and Lubricants*, Elsevier [1].)

TABLE 1.1. The Coefficient of Friction for Different Pairs of Metallic Materials in Dry State [10].

	W	Mo	Cr	Co	Ni	Fe	Nb	Pt	Zr	Ti	Cu	Au	Ag	Al	Zn	Mg	Cd	Sn	Pb	In
In	1.06	.73	.70	.68	.59	.64	.67	.79	.70	.60	.67	.67	.82	.90	1.17	1.52	.74	.81	.93	1.46
Pb	.41	.65	.53	.55	.60	.54	.51	.58	.76	.88	.64	.61	.73	.68	.70	.53	.66	.84	.90	
Sn	.43	.61	.52	.51	.55	.55	.55	.72	.55	.56	.53	.54	.62	.60	.63	.52	.67	.74		
Cd	.44	.58	.58	.52	.47	.52	.56	.59	.50	.55	.49	.49	.59	.48	.58	.55	.79			
Mg	.58	.51	.52	.54	.52	.51	.49	.51	.57	.55	.55	.53	.55	.55	.49	.69				
Zn	.51	.53	.55	.47	.56	.55	.50	.64	.44	.56	.56	.47	.58	.58	.75					
Al	.56	.50	.55	.43	.52	.54	.50	.62	.52	.54	.53	.54	.57	.57						
Ag	.47	.46	.45	.40	.46	.49	.52	.58	.45	.54	.48	.53	.50							
Au	.46	.42	.50	.42	.54	.47	.50	.50	.46	.52	.54	.49								
Cu	.41	.48	.46	.44	.49	.50	.49	.59	.51	.47	.55									
Ti	.56	.44	.54	.41	.51	.49	.51	.66	.57	.55										
Zr	.47	.44	.43	.40	.44	.52	.56	.52	.63											
Pt	.57	.59	.53	.54	.64	.51	.57	.55												
Nb	.46	.47	.54	.42	.47	.46	.46													
Fe	.47	.46	.48	.41	.47	.51														
Ni	.45	.50	.59	.43	.50															
Co	.48	.40	.41	.56																
Cr	.49	.44	.46																	
Mo	.51	.44																		
W	.51																			

4

deformation, which begins when a normal stress exceeds the compressive yield value Y of steel. On the other hand, copper initiates creep-like flow, similar to a lubricant film, when the stress becomes larger than the shear yield value k. The coefficient of friction between these substances was derived as:

$$\mu = k/Y \qquad (1.2)$$

The predicted value was found to be in good agreement with that determined by the friction measurements [11]. However, deviation often appears, since there are other factors influencing the frictional force. As an example, let us consider a hard steel ball (radius r) sliding over a hard steel plate, with soft copper film placed in between (Figure 1.4). When the ball begins to move with a heavy load W, the ball will penetrate into the soft copper and then necessarily displace the metal ahead of the ball. If the substrate steel plate is too hard to be plastically deformed, the contacting area A' can be expressed as Equation (1.3) as a result of depression.

$$A' = cW^{2/3} \qquad (1.3)$$

where c is a constant.

The theory of elastic deformation was first introduced by Hertz, and a number of elastic formulae have been derived from this theory. According to his formula, for instance, the constant c may be derived as:

$$c = 1.2\pi[\, r/2(1/E + 1/E')]^{2/3} \qquad (1.4)$$

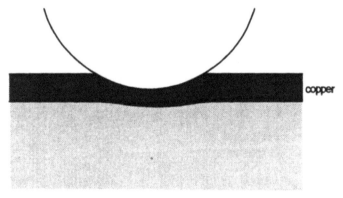

FIGURE 1.4. A hard ball slides over a steel plate lubricated by a copper film. (Reproduced from *Lubrication and Lubricants*, Elsevier [1].)

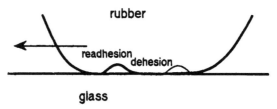

FIGURE 1.5. Schematic diagram showing how a wave of detachment travels through the contact zone.

where E and E' are the Young's modulus of the steel plate and the steel ball, respectively. In such a case, the apparent μ is no longer constant with different loads.

1.1.2 Polymers

When a thermoplastic polymer or an elastomer slides over a relatively smooth and hard surface or over a soft material surface, complicated factors other than those for the rigid materials may arise [12]. A schematic diagram is depicted in Figure 1.5 as an example to show how a wave of detachment travels through the contact zone when a soft rubber moves over a smooth glass surface. The dehesion and readhesion at the rubber/glass interface enables the relative motion to occur between the rubber and the glass without interfacial sliding actually taking place [13]. When an elastomer or a polymer is brought into contact with another material, various kinds of adhesive force are involved at the contacting interface, arising from electrostatic, van der Waals, or dipole interactions between the groups existing in the surface region.

When a hard hemisphere material is brought into contact with a flat rubber plate and allowed to slide over the plate, air will be squeezed into the advancing contacting interface and will move to the rear part at a high speed. This phenomenon is known as a Schallamach wave and has been studied intensively by Roberts and Thomas [14]. If the Schallamach wave moves across a constant length l at a velocity ω, while the rubber, subjected to a sliding force F, is traveling on a hard material at a speed V, and if the energy

TABLE 1.2. *Generation of Schallamach Wave and Surface Energy.*

Sliding Speed V(cm·s^{-1})	Velocity of Schallamach Wave ω(cm·s^{-1})	ω/l (s^{-1})	Natural Rubber-Glass $\gamma\omega$ (erg cm^{-2})	F Calculated Value (dyn cm^{-2})
0.024	0.86	5.6	6.7×10^3	1.5×10^6
0.043	1.07	11.7	8.0×10^3	2.17×10^6
0.093	1.45	23.6	8.4×10^3	2.12×10^6

factor is simplified to express only the peeling-off energy between the two solid materials, the following relation is given:

$$F = \gamma\omega/lV \qquad (1.5)$$

where γ represents the surface energy between the rubber and the hard solid material. Equation (1.5) was derived by taking into account the energy balance between the work done on the rubber and the energy loss associated with the Schallamach wave. The work done in a time dt is given by $FVdt$, and the energy loss should be given by $(\gamma\omega/l)dt$. The frictional force calculated from Equation (1.5) is given in Table 1.2, which shows good agreement between the experimental and the calculated results [14].

Grosch [15] studied the friction of rubber sliding over a glass surface at varying temperatures and sliding speeds below 10 mm/s. Higher velocities produced frictional heating. It was found that the μ-velocity relation could be given by a single master curve, when the velocity axis was displaced by a factor determined from the viscoelastic properties of the rubber using the well-known Williams, Landell, and Ferry (WLF) transform (Figure 1.6).

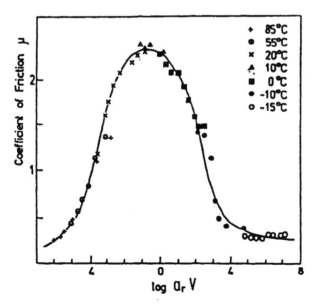

FIGURE 1.6. Coefficient of friction for rubber sliding over a glass surface possessing a slightly irregular topography. The experiments were carried out at various temperatures from $-15\,°C$ to $+85\,°C$ and at sliding speeds ranging from 10^{-3} to 10 mm/sec (over this speed range frictional heating may be neglected). The whole of the data may be plotted against a parameter a_TV where a_T is a function of the temperature so that a_TV is a temperature-compensated velocity [15].

Similar results have been obtained by many other workers, but the explanation of the relations between the adhesion factors and the frictional force of the elastic materials has been far from satisfactory, although macroscopic concepts of adhesion, shearing, and tearing at the interface have been proposed [16,17].

1.2 LUBRICATION

Lubrication takes place through at least three different modes: boundary, fluid-film, and solid-film lubrication. Each of the characteristics is schematically illustrated in Figure 1.7. Fluid-film lubrication includes hydrodynamic lubrication [Figure 1.7(b)] and the hydrostatic lubrication as illustrated in Figure 1.1. The differentiation among various lubrication

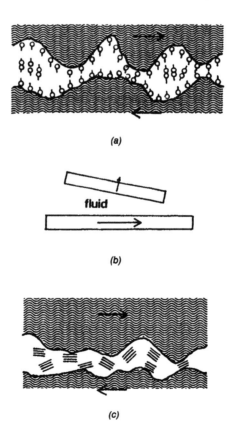

(a)

fluid

(b)

(c)

FIGURE 1.7. Schematic representation of three different modes of lubrication: (a) boundary lubrication; (b) fluid-film lubrication; (c) solid-film lubrication.

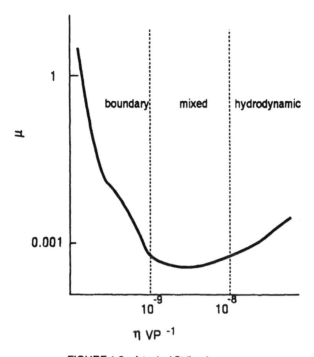

FIGURE 1.8. A typical Stribeck curve.

modes other than the hydrostatic can be made based on the Stribeck curve shown in Figure 1.8. The variation of μ is plotted as a function of a nondimensional parameter $\eta V P^{-1}$, where η is the viscosity of the lubricant, V is the sliding velocity, and P is the pressure on the sliding material.

1.2.1 Boundary Lubrication

Boundary lubrication is defined as that in which the sliding surfaces are separated by a very thin film of lubricant, so that the chemical and physical natures of the surfaces and the lubricant are of major importance. The thin film usually contains only one or two layers of lubricant molecules, which are mostly fatty acids or long-chain alcohols. In general, the lubricants applied for the boundary lubrication become more effective as their chain length increases for each homologous series of chemical structure. When compared at the same chain length, alcohols are less effective than fatty acids, but more effective than paraffins. The effectiveness of organic boundary lubricants is known to correlate closely with the nature of the interfacial solid surfaces. For instance, an oxidized metal surface commonly interacts strongly with surfactants employed as the lubricant, resulting in formation of a good boundary lubricant film up to elevated temperatures. On the other hand, a nonoxidized metal (e.g., platinum or gold) shows no chemisorption or reactions with any lubricants used. In other words, oxidized metals react easily with fatty acids to form soaps, which can maintain the solid state up to considerably high temperatures, while fatty alcohols have lower melting points than their derived acids.

In the boundary lubrication, solid-phase lubricant molecules existing at the interfacial region are generally more effective than liquid-phase lubricants. Therefore, lubricants are sufficiently effective so long as they remain solid below their melting point temperature. However, local heating may easily occur (the so-called hot spot) under a continuous frictional movement, especially at high speeds, resulting in low lubrication. The compounds employed as supplemental reagents for affording good lubrication even at high temperatures are known as "extreme-pressure additives" (E.P. additives). The term "pressure" may not be appropriate, and it should perhaps be replaced with the term "temperature." It probably originated from the engineering field, where lubricants in machine equipment were usually employed at high pressures. The most widely utilized E.P. additives are organic compounds containing halogen or sulfur atoms. At high temperatures where hot spots may have resulted from the frictional sliding, the metal surface would have been stained in the course of the reaction with halogen or sulfur atoms in the presence of E.P. additives, leading to surface oxidation.

Practical surfaces are not at all smooth or flat when considered on a molecular scale, as depicted schematically in Figure 1.7. Even finely polished surfaces are known to have surface roughness of about 300 nm on the average. Bowden and Tabor [5] suggested that the local pressure between two metallic materials would be so high that the asperities could deform plastically even at a low external pressure.

When the real contacting area A_r is subjected to the pressure p in the softer of the two specimens in contact – to which the normal load W is applied – the mean shear strength s may approach that of the bulk material. When the frictional force F is expressed as a horizontal component, the following relation can be derived.

$$F = A_r s, \quad W = A_r p \qquad (1.6)$$

μ is constant when calculated from Equation (1.6). Taking into account Equation (1.1), this case is not likely to be different from the simple model as described in the previous section.

$$\mu = F/W = s/p \qquad (1.7)$$

Equation (1.7) indicates that the frictional behavior in the boundary lubrication also conforms well to the Amontons-Coulomb law. However, in most practical cases, theoretical treatments are very difficult because several other complicated factors influence frictional behavior at the interfacial region of two solid materials.

1.2.2 Fluid-Film Lubrication

In contrast to the boundary lubrication, the opposing two solid materials are completely separated by fluid in the case of fluid-film lubrication. As a result, the friction becomes very low and, in addition, wear of the solid materials may be negligibly small. As already shown in Figure 1.1, extremely low friction will be attained if the fluid film is composed of air or other gaseous materials. Since frictional behaviors are governed by the energy consumption or the shear stress of the fluid film, the theoretical treatment is fairly straightforward and well understood from the classical work of Reynolds, which systematically describes the hydrodynamic flow of fluids.

Suppose a bearing component is divided into sectors, each of which has a plane surface inclined to the direction of motion, as illustrated in Figure 1.9. If a viscous fluid is filled between the upper fixed plane and the other lower plane running with a velocity of U, the velocity of fluid is zero at the contacting point against the upper plane and U at the point touching the lower plane, assuming that the fluid moves with the running plane. As the velocity gradient falls linearly from U (running plane) to 0 (the upper fixed plane) in this simple model the mean velocity of the fluid is $U/2$. The flux of fluid between the two planes is constant, irrespective of the difference of gap height, if the following assumptions are made:

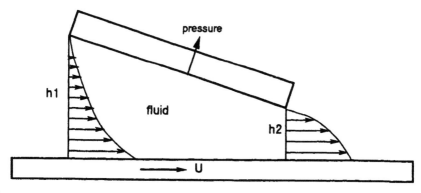

FIGURE 1.9. Velocity profile of a fluid in a converging wedge.

(1) There are no extra forces except gravitational force.
(2) The pressure is constant throughout the thickness of the film.
(3) The velocity of the fluid does not depend on horizontal distance x.
(4) There are no slip phenomena at the boundary.

Under these conditions the following Reynolds equation can be derived:

$$dp/dx(h^3 dp/dx) = 6\eta U dh/dx \qquad (1.8)$$

where dp/dx and η represent the pressure gradient and the viscosity of fluid, respectively.

As a result, when the fluid moves between the two surfaces, the upper plane surface is subjected to different pressures depending on the height (h) of the gap between the two planes. The approximate pressure gradients generated at varying gap heights $(h_1$ and $h_2)$ are shown in Figure 1.10. Integration of Equation (1.8) gives

$$dp/dx = 6U\eta\{(h - h^*)/h^3\} \qquad (1.9)$$

where h^* is the gap height at $dp/dx = 0$. Further integration gives

$$p = 6U\eta \int (h - h^*)/h^3 dx + C \qquad (1.10)$$

where C is a second constant of integration. Equations (1.9) and (1.10) demonstrate only the one-dimensional form of the Reynolds formula, but

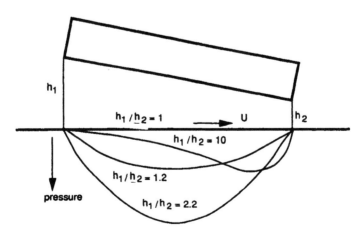

FIGURE 1.10. Pressure gradients appearing on the upper fixed plane at different ratios of h_1 to h_2.

two- and three-dimensional forms can be readily derived. However, solutions for different types of lubrication are so complicated that the practical calculation is usually performed with the aid of a computer. The significance of the Reynolds equation is to demonstrate that the flow of viscous fluid causes an increase in pressure between the surfaces, resulting in a lifting force on the upper plate. If a wide cylindrical bearing is filled with a liquid-oil lubricant, the pressure becomes higher at the contacting area as the shaft runs faster. Thus the shaft will be raised to reduce the convergence until the shaft runs almost concentrically in the bearing, with the load supported entirely by the liquid. At this point there is no substantial contact between the surfaces, leading to very low friction.

1.2.3 Solid-Film Lubrication

As described earlier, the use of E.P. additives is required in the boundary lubrication at high temperatures, to form a solid film. E.P. additives reduce friction more effectively if the solid has a lamellar crystal structure and is firmly attached to the substrate surfaces. Compounds widely used for this purpose include molybdenum disulfide (MoS_2), graphite, boron nitride (BN), titanium iodide (TiI_2), tungsten sulfide (WS_2), titanium sulfide (TiS_2), and so forth. In addition, many other solid materials can be used as solid-film lubricants, including metal powders and polymer films. Although these solid films do not lead to lower friction than fluid films, the solid films are generally independent of velocity and temperature. Furthermore, they can be used for slow and reciprocal motion with heavy weights. Polytetrafluoroethylene (PTFE) is a commonly used solid-film lubricant, which exhibits uniquely low friction under almost any environmental conditions because of its inert property to chemical reagents and its high temperature resistance. The drawback of this polymer is the creep that occurs under a large load, but this is a common phenomenon of most polymeric materials. These solid-film lubricants would also be very effective in such biomedical applications as surgical gloves, sutures, and artificial joints. Nontoxicity toward the human body is, however, a particular requirement for lubricants to be used in these applications. Such lubricants will be described in detail in Chapter 4.

1.3 WEAR

Wear occurs wherever frictional motion takes place at an interfacial region, and lubricity or frictional force greatly influences the wear on both the sliding material and the opposing surface. Although wear is important in lubrication systems, its nature has been poorly understood from a theoretical standpoint, due to the difficulty in conducting experimental measurements with good reproducibility and in constructing analytical means to determine very small amounts of wear particles. In addition, the materials worn away from the surfaces often readhere to both the sliding substrate and the opposing surface, and the quantitative determination often requires transfer of such fragments in quite a short period of time. Furthermore, it is extremely difficult to analyze the wear results, because they are influenced by numerous factors such as temperature, velocity of frictional motion, the nature of the sliding material and the countersurface, and the shapes of the surfaces.

Fatigue, abrasion, and adhesion also play an important role in the wear of polymeric materials. Lancaster [18] studied the relation between the wear rate and the work to produce wear fragments for various polymeric materials. In Figure 1.11, the wear rate of polymers is plotted against the degree of roughness of the polymer surfaces. The breaking stress (S) and the elongation at break (ϵ) are readily determined using a conventional tensile testing machine, and their product ($S\epsilon$) is proportional to the energy up to failure. As is seen in Figure 1.12, where the wear rate of polymers

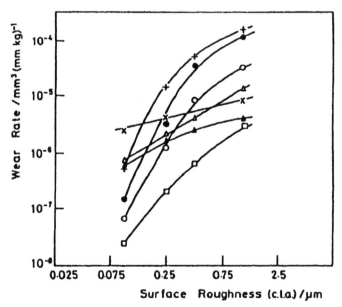

FIGURE 1.11. Variation of wear rate with surface roughness for polymers during single traversals over mild steel rings (from Lancaster [18]): + —polystyrene, ● —PMMA, ○ —acetal homopolymer, △—polypropylene, × —PTFE, ▲—polyethylene, □—nylon 6.6.

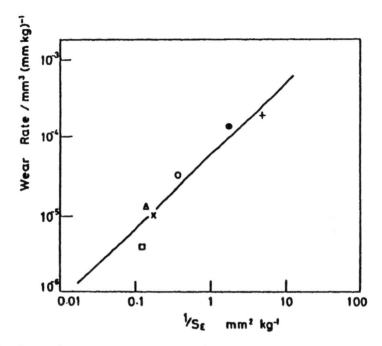

FIGURE 1.12. Correlation between wear rates of polymers and the reciprocal of the energy parameter S_ϵ. The wear rates refer to single traversals over a rough steel surface (c.l.a. = 1.2 μm) and the results show that all the data of Figure 1.11 lie close to a single curve [18].

was plotted against the reciprocal of the energy parameter S_ϵ, there is a good correlation between these values. This finding suggests that the wear of polymeric materials is closely related to such a work function.

1.4 REFERENCES

1. Braithwaite, E. R., ed. 1967. *Lubrication and Lubricants.* Elsevier.
2. Cameron, A. 1981. *Basic Lubrication Theory.* Amsterdam: Ellis Horwood Publishers.
3. Moore, D. F. 1972. *The Friction and Lubrication of Elastomers.* Oxford: Pergamon.
4. Lee, L.-H., ed. 1975. *Advances in Polymer Friction and Wear.* New York: Plenum.
5. Bowden, E. P. and D. Tabor. In *The Friction and Lubrication of Solids, Part I* (1950); and *Part II* (1964). London: Oxford University Press.
6. Bhushan, B. and B. K. Gupta. 1991. *Handbook of Tribology: Materials, Coatings, and Surface Treatments.* New York: McGraw-Hill.
7. Amontons, S. G. 1699. *Hist. Acad. Roy. Sci.*, p. 206.
8. Coulomb, C. A. 1785. *Mem. Math. Phys. R.S.*, p. 161.

9. Whitehead, J. R. 1950. "Surface Deformation and Friction of Metals at Light Loads," *Proc. Roy. Soc.*, A201:109-124.

10. Rabinowicz, E. 1971. *ASLE Trans.*, 14:198.

11. Male, A. T. and M. G. Cockcroft. 1964. "A Method for Determination of Coefficient of Friction during Plastic Deformation," *J. Inst. Metals*, 93:38-46.

12. Briscoe, B. J. and D. Tabor. 1978. "Friction and Wear of Polymers," in *Polymer Surfaces*, D. T. Clark and W. J. Feast, eds. New York: John Wiley, pp. 1-23.

13. Schallamach, A. 1971. "How Does Rubber Slide?" *Wear*, 17:301-312.

14. Roberts, A. D. and A. G. Thomas. 1975. *Wear*, 33:45.

15. Grosch, K. A. 1963. "The Relation between the Friction and Viscoelastic Properties of Rubber," *Proc. Roy. Soc.*, A274:21-39.

16. Ludema, K. C. and D. Tabor. 1966. "Friction and Viscoelastic Properties of Polymeric Solids," *Wear*, 329-348.

17. Tabor, D. 1975. "Friction, Adhesion and Boundary Lubrication of Polymers," in *Advances in Polymer Friction and Wear*. New York: Plenum, pp. 5-30.

18. Lancaster, J. K., 1972. "Friction and Wear," Chapter 14 in *Polymer Science, A Materials Science Handbook*, A. D. Jenkins, ed. North-Holland Pub.

CHAPTER 2

Test Methods for Lubricated Polymer Surfaces

A variety of methods have been proposed for testing the friction and wear of polymeric materials. The principle for determining μ is essentially the same among the different methods, that is, it is done by measuring the ratio of the frictional force to the normal load applied on the specimen in the frictional motion, as represented in Equation (1.1) in Chapter 1.

2.1 METHODS FOR DETERMINING COEFFICIENTS OF FRICTION

ASTM D 1894-90 describes the standard test method for determining the static and kinetic coefficient of friction of plastic film and sheeting. According to this method, the size of the test specimen is $4.5'' \times 4.5''$, and that of the test surface is $10'' \times 5''$. There are five possible contact arrangements between the test surface and the test specimen, as depicted in Figure 2.1. Of these different arrangements, (c) and (d) require the use of a low-friction pulley, while (a), (b), and (e) do not. Only the (d) type arrangement needs no motor drive. In each of these cases the friction drag should be measured at $23 \pm 2°C$ and $50 \pm 5\%$ relative humidity with an accuracy of $\pm 5\%$.

2.2 METHODS FOR DETERMINING WEAR

Wear is abrasion (weight loss) of a material resulting from the frictional action upon the material surface. British Standard (BS) defines four methods of testing vulcanized rubber, as shown in Figure 2.2. The following is a brief description of the principles of these four different methods.

- Method A [Figure 2.2 (a)]: The flat end of a cylindrical test piece is abraded against the surface of a rotating drum covered with an abrasive cloth, while the test piece is traversed from one end of the drum to the other to reduce contamination of the cloth.

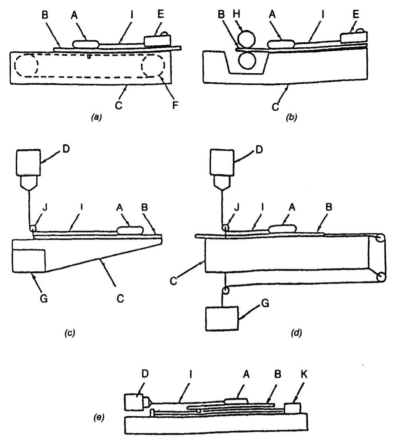

FIGURE 2.1. Test methods by ASTM. A: slider, B: test surface, C: plate, D: gauge, E: spring gauge, F: belt drive, G: crosshead, H: roll, I: nylon monofilament, J: low-friction pulley, K: motor.

- Method B [Figure 2.2 (b)]: A rotating test piece in the form of a disc is abraded against an abrasive wheel under a specified load. The planes of the test piece and the wheel are inclined at an angle. A duct is applied, if required, during the testing in order to prevent clogging of the abrasive wheel with rubber debris.
- Method C [Figure 2.2 (c)]: A flat test piece is abraded against a rotating abrasive disc under a specified load.
- Method D [Figure 2.2 (d)]: A flat test piece is subjected to rotary slip between itself and a pair of abrasive wheels. The test is carried out using one of a range of wheel loadings and abrasive surfaces.

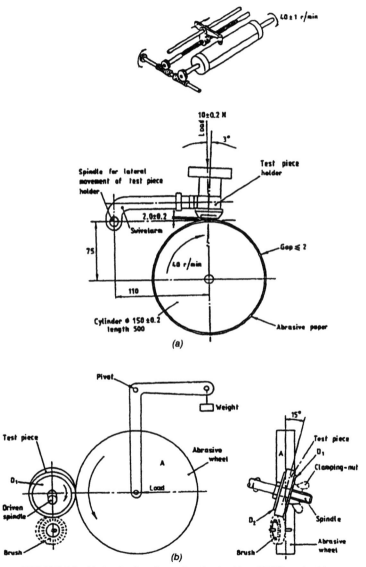

FIGURE 2.2. Methods of testing vulcanized rubber (BSI Standards).

19

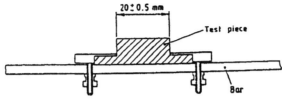

Dimension is in millimetres.

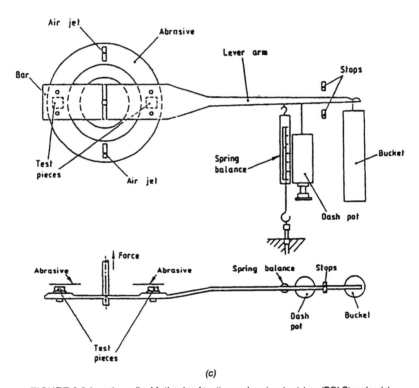

(c)

FIGURE 2.2 (continued). Methods of testing vulcanized rubber (BSI Standards).

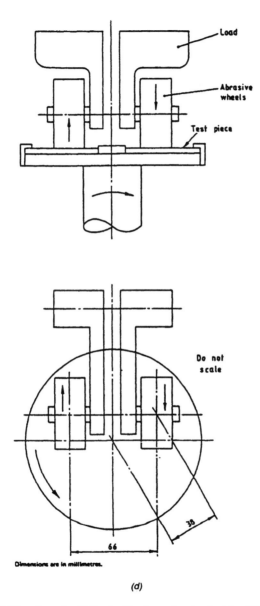

Load

Abrasive
wheels

Test piece

Do not
scale

38

66

Dimensions are in millimetres.

(d)

FIGURE 2.2 (continued). Methods of testing vulcanized rubber (BSI Standards).

21

2.3 METHODS USED IN PRACTICE

2.3.1 Modified ASTM Method

Although the ASTM standard in Section 2.1 defines how to determine the coefficient of friction with a rather simple principle, it has inherent problems, such as requirement of a large specimen, noncompact equipment, etc. Bowden and Leben [1] devised an original apparatus for friction measurements. Figure 2.3 shows the Bowden-Leben friction testing apparatus, in which, for instance, a hard material (e.g., steel ball "B") is moved over flat test specimen "A". B is loaded with a ring-shaped spring "F". As a plate "C" moves horizontally in the direction of the arrow, mirror "M" would move according to the frictional force between the test piece and the hard material. The frictional force can be calculated from the displacement of the mirror.

Triolo and Andrade modified ASTM-27 D 1894-63 to determine the frictional force of various polymer surfaces in water, isotonic saline solu-

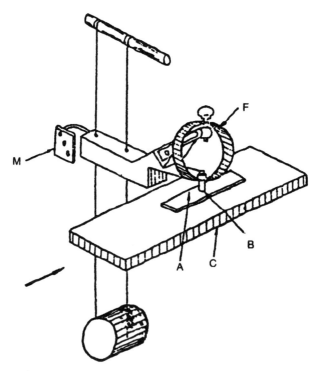

FIGURE 2.3. Bowden-Leben friction testing apparatus. A—test specimen; B—steel ball; C—plate; F—leaf spring; M—mirror.

tion, and blood plasma [2]. The polymers tested included silicone rubber, polyethylene, and fluorinated ethylene propylene copolymer. The frictional force of these polymers was measured before and after modifying their surfaces with radio frequency glow discharge. The apparatus consisted of a constant temperature bath, a trough into which the test pieces and the friction environment (air, water, blood plasma, etc.) were placed, a tensile testing machine (Instron Model TM5) with a 500 g load cell set to move one surface past the other, a low-friction pulley to transmit the vertical force supplied by the Instron to the horizontal motion of the test pieces across one another, a sled on which the test pieces were mounted for testing, and a piece of silk fishing line to connect the sled to the load cell of the Instron. Figure 2.4 shows typical representations of friction profiles in aqueous environments. The profiles vary significantly from polymer to polymer. Some are very smooth, but others have frequent, high-amplitude spikes, while others have low-frequency and low-amplitude spikes. Triolo and Andrade discussed the mechanisms responsible for the frictional behavior, classifying the profiles into the following five categories (see Figure 2.4):

- Type A: The force versus slider traveling distance curve has many high-amplitude and low-frequency spikes with sharp peaks and steep slopes.
- Type B: High-amplitude and low-frequency spikes are separated by sections of lower force. Peaks are sharp and fall rapidly, but rise slowly.
- Type C: Low-amplitude and high-frequency spikes predominate.
- Type D: The profile curve is very smooth, usually with a slight spike at the beginning of the motion, followed by a nearly constant force.
- Type E: A relatively smooth curve is produced, but the force required to keep the slider moving increases gradually with time, passing through a maximum, and then decreases again.

A method similar to Figure 2.1(c) was further utilized for determining the coefficient of friction between the surface of a polymer film grafted with various hydrophilic polymers and a glass plate under water [3]. A brief schematic representation of the apparatus is shown in Figure 2.5. If frictional force is extremely small, such as several grams of force due to the grafted hydrophilic polymer, other additional forces – such as those arising from the pulley or the surface tension between the slider and the water – cannot be neglected, as they are with the ASTM method, which uses a relatively large contacting area and a large load. The values of the coefficient of friction were calculated from the slope of the linear dependency, obtained by plotting the frictional force against a load put on the specimen. Representative plots for polypropylene films are shown in Figure 2.6.

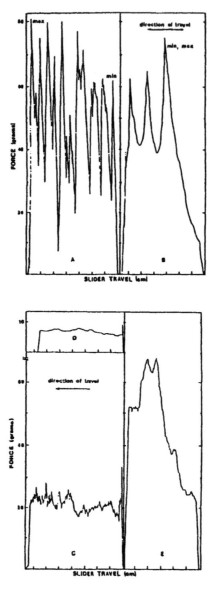

FIGURE 2.4. Friction profiles of polymers in aqueous environments. Type A appeared to be indicative of "stick-slip" friction. Type B appeared to be intermediate between types A and C. Energy was released from the polymer more slowly than it was dissipated (accounting for the rise in the peak) until a critical level was reached and the energy was suddenly released. "Min" is the amount of force recorded as the first peak value. "Max" is the maximum force value obtained. Type C is similar to type A behavior except the surfaces were more rigid and/or the interfacial forces were not as great and/or the surface region of one of the two polymers was weak. Type D behavior is that which we would expect from frictional phenomena. An initially high resistance to friction indicative of static friction is followed by a region of nearly constant resistance indicative of dynamic friction. Type E behavior appears to be intermediate between types A and D [2].

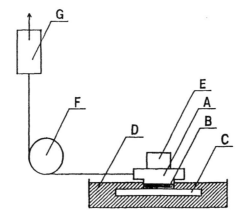

FIGURE 2.5. Schematic of the apparatus used for determining the coefficient of friction of a hydrated surface against a glass plate in water: A—slider; B—sample film; C—glass plate; D—distilled water; E—load; F—pulley; G—100 g load cell.

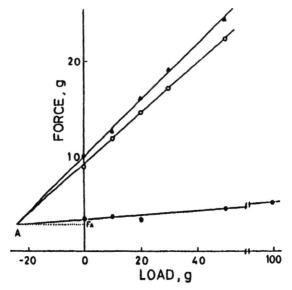

FIGURE 2.6. Plots of the average frictional force against the load put on the slider for the starting PP film (○) and the PAAm-grafted PP film (●). The symbol (△) shows the static friction for the starting PP film [3].

25

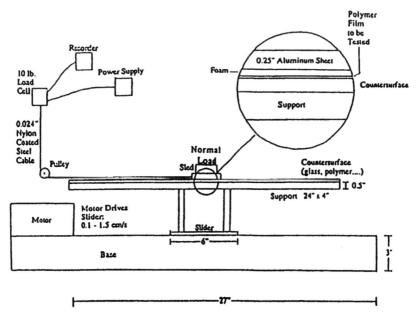

FIGURE 2.7. Instrument built to measure the coefficient of friction of polymer films. (Source: Bee and McCarthy, 1990 [5].)

Graiver et al. [4] also determined the apparent friction coefficient for a PVA hydrogel-coated catheter material using the similar combination of apparatus as shown in Figure 2.5.

Another form of the ASTM arrangement of sliding system [the moving countersurface, Figure 2.1, (b) and (d) types] was used by Bee and McCarthy [5], as shown in Figure 2.7. Friction measurements were conducted by sliding 2.0″ × 2.5″ pieces of the film under study over a PET countersurface under a 300 g normal load at a sliding speed of 0.10 cm/s for a distance of 25 to 30 cm. They found that the above condition was the optimum condition for high reproducibility of results. The coefficient of friction was calculated by dividing the measured frictional force by the applied normal force.

2.3.2 Sliding Angle

Perhaps the simplest method for determining the coefficient of static friction is to measure the friction angle at which the specimen starts sliding. For this purpose, the test surface is set horizontally. A test specimen is placed upon it, and then the surface is gradually inclined until the test specimen begins to slide. The sliding angle θ, which is the angle of the test surface slope against the horizontal plane, is related to the coefficient of friction μ by:

$$\mu = \tan \theta \qquad (2.1)$$

The main drawbacks of this method are that it is not applicable to dynamic friction, and that it is difficult to read with accuracy the sliding angle for an extremely low friction specimen. Moreover, polymers having high elasticity often adhere strongly to the test surface and, as a result, the test specimen does not slip down even at an angle close to 90°.

Cho and Yasuda [6] determined the μ value of plasma-polymerized polymers by measuring the sliding angle of a steel slider (2 cm × 2 cm × 0.2 cm) on a microslide glass coated with plasma polymers. They employed hydrocarbons, halocarbons, and organosilicones as the plasma polymers. Most of these plasma polymers had relatively low friction coefficients although none of them was as low as that of Teflon (0.05−0.1).

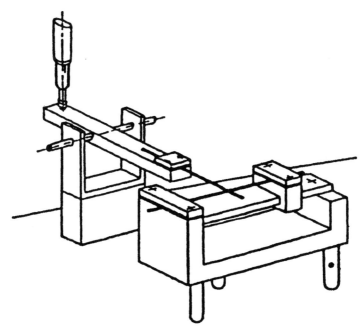

FIGURE 2.8. Friction testing apparatus for monofilament fibers [7].

2.3.3 Fibers

The surface lubricity of yarns, threads, and monofilament fibers is difficult to determine due to their small contacting area. Pascoe and Tabor [7] attempted to measure the frictional force between two fibers using an apparatus shown in Figure 2.8. Friction measurements for surgical sutures will be described in more detail in the following section.

2.4 MEDICAL DEVICES

It is often difficult to determine the μ values of biomedical materials due to their complicated shapes. In addition, these biomaterials may exhibit frictional profiles different from those of *in vitro* tests when inserted into the body, due to various interactions with proteins and cells present in the body.

2.4.1 Catheter

Nagaoka and Akashi [8] measured the friction of a catheter surface modified with a cross-linked $100 \mu m$ thick collagen film. The catheter was cut to a length of 5 cm and fixed on a glass plate as shown in Figure 2.9. The catheter was wetted with physiological saline solution, and a 100 g weight whose bottom was covered with the collagen film and wetted with the same fluid, was placed on the catheter. The measurement of the sliding angle was performed by alternating the specimen and the opposing test surface.

A measuring system depicted in Figures 2.10 and 2.11 was used to study the slipperiness and the lubricity of catheters [9 − 11].

2.4.2 Joint

The friction in natural and artificial joints is much more difficult to study because the shape of joints is complicated and not uniform. Moreover, the amount of synovial fluid filling human joints is very small, making it difficult to obtain for the frictional experiment. Therefore, the frictional *in vitro* measurement is often performed using artificial synovial fluids and, for instance, a pendulum method, as depicted in Figure 2.12. The principle of this method is based on a determination of the decrement of oscillatory movement. The frictional moment M resulting from the frictional motion between the two joint articles during one cycle is obtained from the angular loss $\Delta\theta$ derived from the average difference between the angle of the $(n-1)$th cycle and the nth cycle of oscillation:

$$M = 1/4 \, l\Delta\theta w \qquad (2.2)$$

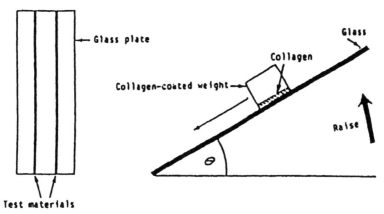

FIGURE 2.9. Method of determining surface friction coefficients of collagen-coated material.

where *l* and *w* are the length and the weight of pendulum, respectively. Hence, the μ value is calculated in Equation (2.3):

$$\mu = M/r\,W = l/r\,\Delta\theta/4 \qquad (2.3)$$

where *r* is the radius of the sphere to be tested. Although this method has

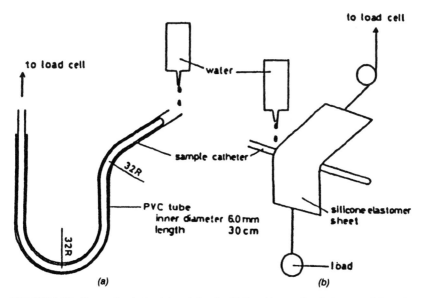

FIGURE 2.10. The methods for determining the frictional force of catheter materials under wet conditions against a plasticized PVC tube (a) and a silicone elastomer sheet (b).

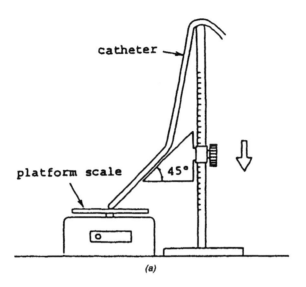

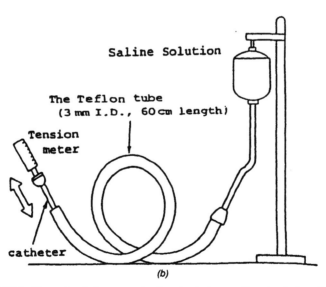

FIGURE 2.11. The methods for measuring the physical properties of catheters. (a) Stiffness and elasticity of the catheters. (b) Measurement of the surface coefficient of friction of the catheters.

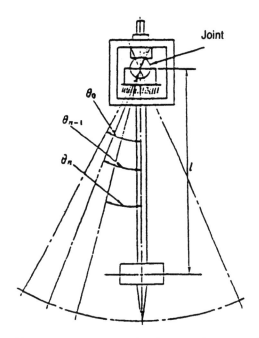

FIGURE 2.12. A pendulum method for measuring joint friction.

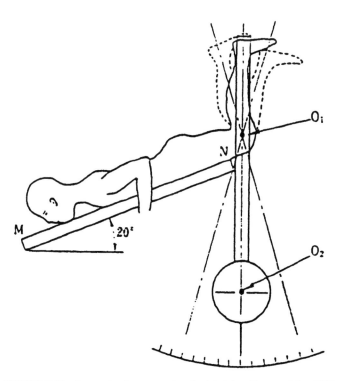

FIGURE 2.13. Apparatus for measuring the lubricity of human joints [12].

several drawbacks [such as the μ value calculated from Equation (2.3) is neither the static (μ_s) nor the kinetic friction coefficient (μ_k), but only a flattened coefficient of friction based on one cycle of oscillation under a restricted velocity range of oscillation], it is often utilized to estimate the lubricity of a joint. Sasada and Maesawa [12] studied the lubrication of human joints using the technique illustrated in Figure 2.13. In such a case, r in Equation (2.3) is estimated from the X-ray photograph of a patient.

Komoto et al. [13] prepared an artificial joint model from PVA which had a hydrogel structure similar to that of native articular cartilage. This model was prepared by dissolving PVA in water, and directly grafting it by gamma-ray irradiation onto the surface of a high-density PE disk. Friction measurement was carried out by rotating the specimen disk in water against a steel ball of 4.26 mm diameter at a speed of 18.8 cm/s. The obtained μ value was as low as 0.02 in the beginning of the friction experiment. However, μ was found to gradually increase to almost the same value as that of the steel/HDPE system (0.06), as the apparently grafted PVA layer was readily worn away by friction.

Oka et al. [14] also studied the dynamic change of the gap between a glass plate and a natural joint or an artificial articular cartilage made of ultrahigh molecular weight PE (UHMWPE) or PVA hydrogel as well as damping

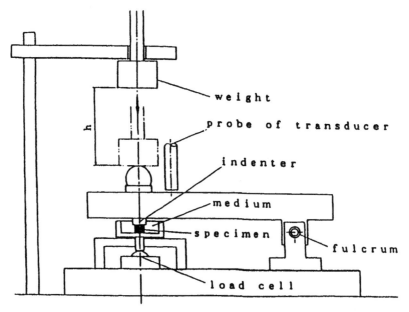

FIGURE 2.14. Diagram of the apparatus used for damping measurement. A weight (27 N) was dropped from 10 mm height on the specimen in fluid. The stress transmitted is measured by a load cell set up under the specimen [14].

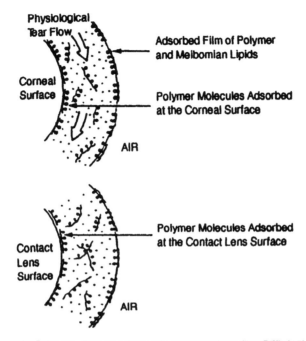

FIGURE 2.15. Role of polymers in precorneal tear film (PTF) [15].

effects of various prosthetic materials to confirm the existence of fluid-film lubrication in the natural joint. The apparatus they employed is depicted in Figure 2.14. In the case of UHMWPE, the fluid pressure in the gap became zero soon after loading and the gap width became very narrow. On the other hand, the fluid pressure for the PVA hydrogel remained over 1 MPa for 1 h after loading, and the gap width was maintained at the thickness of natural articular cartilage. It was concluded that UHMWPE is not a good material from the viewpoint of fluid-film lubrication, whereas the PVA hydrogel appears to be a promising and interesting joint material.

2.4.3 Contact Lens

Kalachandra and Shah [15] measured the coefficient of friction between two PMMAs in the presence of various ophthalmic polymer solutions. The role of polymer solutions as precorneal tear film (PTF) is schematically depicted in Figure 2.15. The surface tension and the contact angle on PMMA were found to be independent of the viscosity of the polymer solutions used, and the contact lens system operated as a system of boundary lubrication. The results are given in Table 2.1, where it is seen that PVA has the lowest coefficient of friction. The effects of these polymer solutions on μ may depend on the chemical structure, conformation, and adsorption

*TABLE 2.1. Surface Chemical and Lubrication Parameters
of Various Ophthalmic Polymer Solutions [15].*

Ophthalmic Polymer	Coefficient of Friction* μ	Surface Tension dynes/cm γ	Contact Angle θ of the Solution on Clean PMMA in Degrees
1. Carboxyl Methyl Hydroxy Ethyl Cellulose	0.234	50.0	42
2. Hydroxy Propyl Methyl Cellulose	0.191	48.0	40
3. Hydroxy Butyl Methyl Cellulose	0.187	64.7	36
4. Carboxy Methyl Cellulose	0.162	62.5	46
5. Chondroitin Sulfate	0.152	59.0	46
6. Dextran	0.147	55.0	45
7. Methyl Cellulose	0.131	58.0	42
8. Poly(vinyl methyl ether)	0.119	46.0	30
9. Polyvinyl Pyrrolidone	0.117	55.0	33
10. Polyvinyl Alcohol	0.109	46.0	43

*Coefficient of friction was measured at a stylus velocity of 400 mm/sec., vertical load 10 grams and viscosity 20 cp.

characteristics of the polymers as well as the surface characteristics of the sliding surface, load, and speed.

The evaluation of lubricity between a contact lens and a corneal model was performed using the sliding angle method [16]. As depicted in Figure 2.16, a surface-treated contact lens attached to a rod (test sample) is first placed on the outer surface of a corneal model prepared from a PVA hemispheric hydrogel and then the stage carrying the test sample and the hemisphere is gradually inclined. To study the effect of the medium's presence between the corneal model and the lens, different kinds of liquid are added to the interfacial region. The angle θ at which the contact test sample (the total mass M) starts to slide is carefully determined by the scale provided to the stage. The μ value between the contact lens and the opposing surface can be calculated by Equation (2.4).

$$Mgh \sin \theta = r\mu(Mg \cos \theta + a) \tag{2.4}$$

where r, g, and h are the radius of the hemispheric test surface, the gravitational acceleration, and the distance between the centroid of the test sample and the center of the hemisphere, respectively. $Mg \cos \theta$ corresponds to the normal load of the test sample against the hemisphere, while $Mgh \sin \theta$ corresponds to the frictional moment at the beginning of the sliding. The constant a in Equation (2.4) is an additional factor resulting mainly from the surface tension of the liquid film between the two solid materials, which always exists in this experimental arrangement.

It was first shown in 1977 by Katz and co-workers [17,18] that physical contact between the acrylic intraocular lens (IOL) surface and the corneal endothelium was the main cause for corneal endothelial damage during intraocular lens insertion. They demonstrated that there is a direct relationship between the extent of IOL-endothelium contact and the amount of endothelial cell damage through scanning electron microscope studies. They also measured the force of adhesion between IOL materials and the endothelium of excised rabbit corneas using an instrument depicted in Figure 2.17 [19]. The instrument was composed of four main parts:

(1) A cornea-holding device made from transparent PMMA: The excised corneas were inverted, placed on the endothelial surface seeded on the hemispheric ball, and secured at the limbus with a metal ring that was screwed into the plastic.

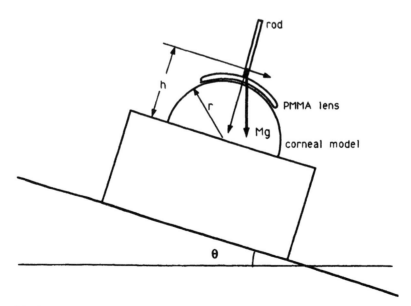

FIGURE 2.16. Scheme for determining the static coefficient of friction between a contact lens and a corneal model.

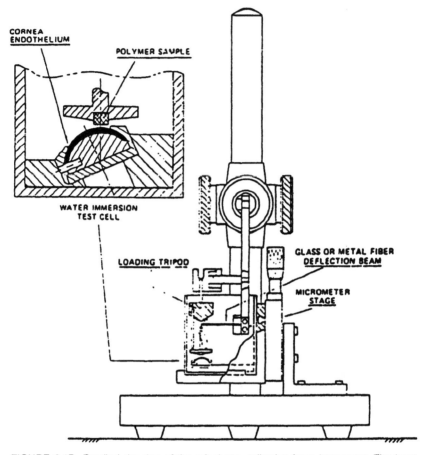

FIGURE 2.17. Detailed drawing of tissue/polymer adhesive force instrument. The inset shows the cornea holding device and the plastic probe [19].

(2) Fiber beam and plastic probe
(3) Pressure device consisting of a tripod supporting a stainless steel piece
(4) Micrometer and cathetometer

The micrometer was used to observe the movement of the edge of the beam through the cathetometer. The difference in scale readings before and after disengagement is proportional to the adhesion force F between the endothelium and the material. The deflection Δd of the beam for small forces is given by

$$\Delta d = 4FL^3/3ER^4 \qquad (2.5)$$

where E, L, and R are the Young's modulus, the length, and the radius of the beam, respectively. The instrument was periodically calibrated using known weights in the range of the measured forces to translate the observed deflection of the beam into the force from the calibration curve. In their instrument, both the deflection beam and the tissue/polymer interface were immersed in water to eliminate the surface tension effect.

2.4.4 Suture

Good handling characteristics with respect to the knot security and tie-down resistance are important for surgical sutures, as is surface lubricity. The tie-down capacity of braided sutures was measured by two different models, as illustrated in Figure 2.18 [20,21]. For both cases, the frictional force between the contacted or the twisted pair of sutures was measured by pulling down the sutures. Tomita et al. [22] attempted to quantify the handling characteristics of sutures through various mechanical tests. They measured the tie-down capacity using a system shown in Figure 2.19. The suture thread to be tested was wound around a sponge tube, tied with a square knot, and placed in the jaws of a testing machine. The sponge tube was wound by the suture as the crosshead was descending, and the resistance force applied to the suture was recorded at a crosshead speed of 30 mm min^{-1}. They also estimated the pull-out resistance from the measurements of thread-thread friction. A modified system similar to that shown in Figure 2.18 was used for the measurement. Two pieces of thread were entwined horizontally and a weight was connected to each of the ends. Another piece

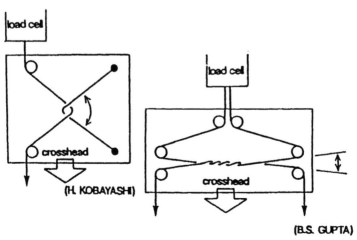

FIGURE 2.18. Conventional friction tests for suture materials.

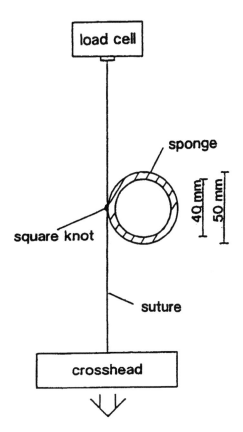

FIGURE 2.19. Tie-down test of sutures.

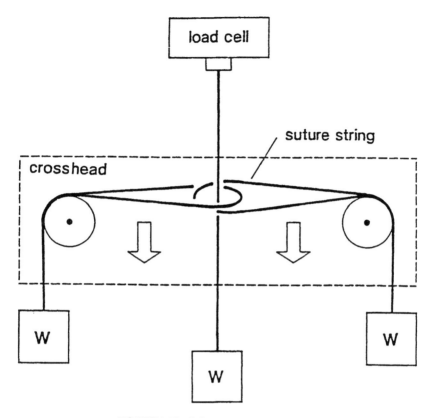

FIGURE 2.20. Pull-out friction test of sutures.

of thread was vertically pulled up through the entwined threads under tension, as depicted in Figure 2.20.

Schlatter [23] described a machine for measuring the friction of a yarn sliding over itself and studied the variation of friction with speed and the slip-stick variety at slow speeds. According to his instrument, it was possible to maintain a desired initial yarn tension through a wide and continuously variable range of yarn speeds.

2.5 REFERENCES

1. Bowden, F. P. and L. Leben. 1938. *Proc. Roy. Soc.*, A169:371.
2. Triolo, P. M. and J. D. Andrade. 1983. "Surface Modification and Characterization of Some Commonly Used Catheter Materials. II. Friction Characterization," *J. Biomed. Mat. Res.*, 17:149–165.

3. Uyama, Y., H. Tadokoro and Y. Ikada. 1990. "Surface Lubrication of Polymer Films by Photoinduced Graft Polymerization," *J. Appl. Polym. Sci.*, 39:489–498.

4. Graiver, D., R. L. Dural and T. Okada. 1993. "Surface Morphology and Friction Coefficient of Various Types of Foley Catheter," *Biomaterials*, 14(6):465–469.

5. Bee, T. G. and T. J. McCarthy. 1990. "Frictional Studies of Surface Modified Polychlorotrifluoroethylene," *Amer. Chem. Soc. Polym. Mater. Sci. Eng.*, 63:94–98.

6. Cho, D. L. and H. Yasuda. 1987. "Tribological Application of Plasma Polymers," *Amer. Chem. Soc., Polym. Mat. Sci. Eng.*, 58:420–423.

7. Pascoe, M. W. and D. Tabor. 1956. "The Friction and Deformation of Polymers," *Proc. Roy. Soc.*, A235:210–224.

8. Nagaoka, S. and R. Akashi. 1990. "Low Friction Hydrophilic Surface for Medical Devices" *J. Bioactive Compatible Polym.*, 5:212–226.

9. Takeuchi, M. and M. Onohara. Japan patent 633866, Jan. 8, 1988.

10. Uyama, Y., H. Tadokoro and Y. Ikada. 1991. "Low-Frictional Catheter Materials by Photo-Induced Graft Polymerization" *Biomaterials*, 12:71–75.

11. Noishiki, Y., Y. Yamane, M. Takahashi, O. Kawanami, Y. Futami, T. Nishikawa, N. Noguchi, S. Nagaoka and Y. Mori. 1987. "Prevention of Thrombosis-Related Complications in Cardiac Catheterization and Angiography Using a Heparinized Catheter (Anthron®)," *Trans. Am. Soc. Artif. Organs*, 33:359–365.

12. Sasada, N. and S. Maesawa. 1973. *Junkatsu*, 18(12):901.

13. Komoto, T., K. Tanaka and S. Hironaka. 1984. "Poly(vinylalcohol)-Grafted Polyethylene as a Model for an Artificial Joint," *Sen-i Gakkaishi*, 40:53–56.

14. Oka, M., T. Noguchi, P. Kumar, K. Ikeuchi, T. Yamamuro, S.-H. Hyon and Y. Ikada. 1990. "Development of an Artificial Articular Cartilage," *Clinical Mater.*, 6:361–381.

15. Kalachandra, S. and D. O. Shah. 1989. "Polymers as Ophthalmic Lubricating Agents," *Mat. Res. Soc. Symp. Proc.*, 110:463–469.

16. Ichijima, E., H. Kobayashi, Y. Ikada, J. Akita and K. Ikeuchi. 1989. "An Estimation Method for the Friction between a Contact Lens and Corneal Modes," *Japanese Soc. Biomater. Prep.*, p. 95.

17. Katz, J., H. E. Kaufman, E. P. Goldberg and J. W. Sheets. 1977. "Prevention of Endothelial Damage from Intraocular Lens Insertion," *Trans. Am. Acad. Ophthalmol. Otolaryngol.*, 83:204–212.

18. Kaufman, J., J. Katz, J. Valenti, J. W. Sheets and E. P. Goldberg. 1977. "Corneal Endothelium Damage with Intraocular Lenses: Contact Adhesion between Surgical Materials and Tissue," *Science*, 198:525–527.

19. Reich, S., M. Levy, A. Meshorer, M. Blumental, M. Yalon, J. W. Sheets and E. P. Goldberg. 1984. "Intraocular-Lens-Endothelial Interface: Adhesive Force Measurements," *J. Biomed. Mater. Res.*, 18:737–744.

20. Kobayashi, H. 1976. "Suture Material," *Geka Mook* (No. 4):1–14.

21. Gupta, B. S. and K. W. Wolf. 1985. "Effect of Suture Material and Construction on Frictional Properties of Sutures," *Surg. Gynecol. Obset.*, 161:12–16.

22. Tomita, N., Y. Ueda, S. Tamai, T. Morihara, K. Ikeuchi and Y. Ikada. In press. *J. Appl. Biomater.*

23. Schlatter, C. U.S. patent 3209589, Oct. 5, 1965.

CHAPTER 3

Lubricious Polymer Surfaces

3.1 CHARACTERISTICS OF POLYMER SURFACES

Until recently, lubrication was widely considered to be applicable to metals and ceramics, but not to polymers, either in practice or theory. This is probably because lubrication always accompanies wear, against which polymers are quite weak, especially under large normal loadings. This poor wear resistance of polymers is due to their low modulus and rigidity compared with metals and ceramics. This difference is primarily due to the molecular structure of these materials. The smallest units of the constituents of metals and ceramics are associated with each other through primary and secondary bonding at high densities, resulting in high melting temperatures, whereas assembly of the repeating units of polymers is constructed through primary bonds only in one dimension, with very weak lateral bonding. As a result, the constituent units of polymers have a much higher mobility than do metals and ceramics, at least at room temperature, even when cross-linked.

The typical surface structures of polymeric materials are schematically represented on a molecular level in Figure 3.1. As is shown, any segment, polar or nonpolar, will be able to change its location as a result of the environmental conditions around the surface region, since the local motion is not restricted as strictly as that of rigid metals and ceramics. However, this fairly high mobile freedom of segments results in poor resistance against wear, since a fraction of soft surface segments may be readily deformed and stripped off from the surface upon being subjected to a high mechanical stress. A higher density of chemical cross-links or a higher degree of crystallinity of polymeric materials would reduce the segmental motion and enhance the mechanical strength, which would then lead to improved wear resistance. Indeed, modern engineering plastics such as nylon, polyacetal, and ultrahigh molecular weight, high-density polyethylene have been used commercially as parts of gears and bearings, where direct sliding between the plastics and the counterparts takes place.

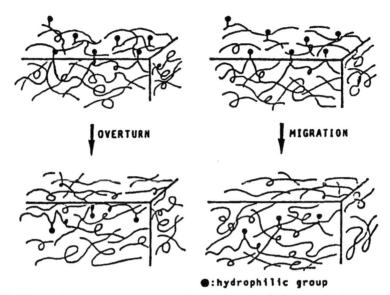

OVERTURN

MIGRATION

●:hydrophilic group

FIGURE 3.1. Schematic representations of the overturn and migration of polymer chains with hydrophilic groups.

It should also be noted that polymer surfaces are often greatly affected by liquids because of the previously mentioned high freedom of segmental motion. This motion allows liquid molecules to enter the interstices of the polymer segments, resulting in polymer swelling even though the density of chemical or physical cross-linking is high. The swelling property, which makes polymers vulnerable to liquids, is quite in contrast to the properties of metals and ceramics.

3.2 FRICTIONAL PROPERTIES OF POLYMERS

Few studies have been done on the friction of polymers, probably because of the relative insignificance of polymer friction in industrial applications. Even if the frictional properties of a polymer surface are excellent in the beginning of its practical use, numerous scratches will be formed on the surface by continuous scraping with another rigid material. This surface degradation will finally lead to failure. In the early 1960s, the socket part of an artificial hip joint was made of polytetrafluoroethylene (PTFE), which had the lowest coefficient of friction among the conventional polymers at that time. However, all the PTFE sockets had to be removed from patients, as extensive wear of PTFE took place against the metal ball counterpart of the artificial joint.

Apart from practical applications, it is of theoretical interest to look at polymer surfaces from the frictional viewpoint. Apparently, no systematic study has been performed to determine the coefficient of friction of a wide variety of organic polymers. One of the reasons for this may be that the frictional property of polymers cannot be absolutely determined, but is greatly dependent on both the counterpart in direct contact and the environmental medium where the polymer-counterpart system is placed. Indeed, neither *The Polymer Handbook, Third Edition*, edited by J. Brandrup and E. H. Immergut (Wiley, 1989), nor *The Encyclopaedia of Polymer Science and Engineering* (Wiley Interscience, 1985) includes a table of coefficients of friction for polymers.

Figure 3.2 shows the coefficients of friction of various polymers determined against a glass plate in pure water. It is apparent that not only very hydrophobic surfaces like PTFE and silicone but also very hydrophilic surfaces like cellulose exhibit very low coefficients of friction in comparison with polar surfaces such as nylon and poly(ethylene terephthalate). We have ascribed this peculiar dependence of the friction on the water contact angle of the polymer surfaces to the work of adhesion of polymers in water [1]. If a polymer surface 1 and a reference surface 2 are in contact and have only dispersive and polar components on their surfaces, the work of adhesion between them in water ($W_{12,w}$) is given by

$$W_{12,w} = 2\{\gamma_1^d [1 - 2(\gamma_1^d \gamma_w^d)^{1/2}/(\gamma_1^d + \gamma_w^d)]\}^{1/2}$$

$$\times \{\gamma_2^d [1 - 2(\gamma_2^d \gamma_w^d)^{1/2}/(\gamma_2^d + \gamma_w^d)]\}^{1/2}$$

$$+ 2\{\gamma_1^p [1 - 2(\gamma_1^p \gamma_w^p)^{1/2}/(\gamma_1^p + \gamma_w^p)]\}^{1/2}$$

$$\times \{\gamma_2^p [1 - 2(\gamma_2^p \gamma_w^p)^{1/2}/(\gamma_2^p + \gamma_w^p)]\}^{1/2} \tag{3.1}$$

where the prefixes d and p denote dispersive and polar components of surface free energy, and the subscripts 1, 2, and w denote the polymer surface, the reference surface, and water, respectively. $W_{12,w}$ values calculated using Equation (3.1) are plotted against the interfacial energy between the polymer surface 1 and water (γ_{1w}) in Figure 3.3. The counterpart reference 2 is assumed in this case to have 30 and 35 erg cm^{-2} as the γ_2^d and γ_2^p, respectively. Figure 3.3 clearly indicates that in both extreme cases where the polymer surface is extremely hydrophilic or hydrophobic, $W_{12,w}$ is quite low, but that it shows a maximum when the polymer surface has a hydrophilic-hydrophobic balance. Since the work of adhesion is directly related to the molecular interactions between the surfaces of both substances, the low $W_{12,w}$ values for the very highly hydrophilic or hydrophobic surface means that no strong interaction is operative on the highly

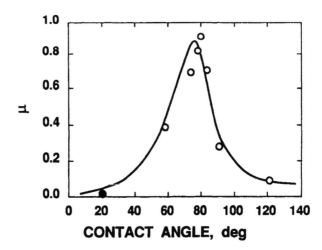

FIGURE 3.2. Relation between the coefficient of friction μ and the water contact angle of films. ○: ungrafted, ●: grafted.

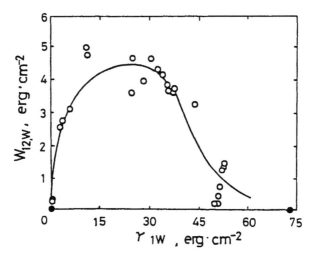

FIGURE 3.3. Work of adhesion between a polymer surface 1 and a reference surface 2 in water ($W_{12,w}$) as a function of the interfacial free energy between water and the polymer surface $1(\gamma_{1w})$ [1].

hydrophilic or hydrophobic surface in water. This seems reasonable because a very highly hydrophobic polymer such as PTFE has no significant dangling molecular force while the strong dangling force of the very highly hydrophilic surface will disappear in a hydrophilic medium such as water. As described in Chapter 1, the liquid film in the fluid-film lubrication is likely to reduce the macroscopic and microscopic interactions between the sliding surfaces.

3.3 LUBRICATION AND SLIPPERINESS

The technical term "lubrication" is used to describe substances that lessen friction between metal parts or objects in relative motion. For this purpose, greases, oils, and solid powders are commonly applied as lubricants, which maintain a fluid film or solid film between the solid rubbing surfaces. On the other hand, many slippery surfaces exist, especially in nature. For instance, an eel has a very slippery surface. Many of the surfaces of the organs in our bodies are also very slippery. The purpose of this slipperiness in nature has not yet been fully clarified, but it is obvious that the slippery surface greatly reduces the frictional resistance occurring when the surface slides on another solid object. In most cases, these slippery surfaces have a large content of water, which acts as lubricant. Another interesting feature of such slippery surfaces is that they prevent other objects from adhering to the surfaces and being damaged. In the following discussion, we do not distinguish rigorously between the words "lubricious" and "slippery," but use them interchangeably.

3.4 SLIPPERY POLYMER SURFACES

As is apparent from Figure 3.2, there are two possible methods for rendering a polymer surface slippery—making the polymer surface either (1) very hydrophobic or (2) very hydrophilic.

3.4.1 Highly Hydrophobic Surfaces

Earlier approaches to providing a low-friction surface were simple application of lubricants such as silicone oil, or coating with a low-friction material such as polyethylene or fluoroplastics. Thomas and Sobel [2] have applied various materials such as silicone, fluorocarbon, and cellulose to polymeric substrates as coatings, in order to decrease surface friction during the process of extrusion or stretching. Their technique involved providing a finite section of hollow extruded thermoplastic tubing that had been cooled to room temperature following an extrusion process. The method further included maintaining the tubing substantially at room temperature conditions and applying a longitudinal stretching force to the tubing sufficient to exceed the elastic limit of the tubing, so that the tubing was deformed from its original dimensions. The tubing may be stretched either before or after its insertion into the hub. As a result of this stretching, the surface lubricity of the tubing was improved to allow it to slide more freely during intravenous insertion.

In 1985, Kent Integrated Scientific System (KISS) proposed polymeric coating for a wide range of applications [3]. The technique developed was to use self-adhesive polymeric coatings that readily adhere in thin films to most solids. KISS claimed that the coating had nonstick surfaces in the wet state, which offered such uses as coatings for boat hulls for drag reduction, dental structures, and cardiovascular prostheses. Moreover, it was claimed that its nonstick, slippery surface acted as an antifouling coating, on which organisms could not readily colonize. Thus, coated glass fiber test plates that had been suspended in the Gulf of Mexico for fifteen months could easily be wiped clean of barnacles and other growths with a bare finger, while uncoated plates could not be cleaned even by scraping with a sharp knife.

3.4.2 Highly Hydrophilic Surfaces

There are numerous patents that disclose the methods for preparing hydrophilic lubricious surfaces. Fan [4] classified the hydrophilic lubricious surface into five categories based on the chemical nature of the resultant surfaces:

(1) Simple coating with hydrophilic polymers
(2) Blending or complexing of hydrophilic polymers
(3) Formation of interpenetrating polymeric networks
(4) Coating with chemically reactive hydrophilic polymers
(5) Surface grafting of hydrophilic monomers

3.4.2.1 PHYSICAL COATING METHODS

Hydrophilic coatings are of major interest among the various methods for preparing lubricious substrates. Indeed, exterior coating of hydrophilic polymers has been the most commonly used technique for imparting a lubricating surface. A dip-coating process may be the simplest and most commonly employed.

Coating methods that use no hydrophilic polymer are also known. For instance, Wong et al. [5] prepared a coating solution consisting of a mixture of curable dimethyl polysiloxane, a cross-linking catalyst (such as dibutyltin diacetate) and an accelerator (such as polysiloxane having pendant amino groups) in toluene. To assure the complete coating, the silicone rubber was dipped twice into the coating mixture and then flash dried in an oven for one minute.

3.4.2.1.1 Hydrophilic Polyurethane

Hydrophilic lubricious surfaces have very often been achieved by coating combinations of polyurethane and hydrophilic, water-soluble poly(vinyl pyrrolidone) (PVP). For example, Althans and Throne [6] applied such a hydrophilic lubricious coating to a razor apparatus. The coating solution consisted of polyurethane dispersion, diacetonalcohol, PVP, and iso-propanol. They reported that the coefficient of friction of coated material was less than 0.25.

Polyurethanes are usually synthesized in a two-step procedure. The first process consists of forming a prepolymer by the reaction of a diisocyanate compound with a low-molecular-weight polyether having dihydroxyl end groups.

$$2(OCN \sim\sim\sim NCO) \ + \ HO \sim\sim\sim\sim\sim OH \rightarrow$$
diisocyanate   polyether

$$OCN - NHCO \sim\sim\sim\sim\sim OCNH - NCO$$
prepolymer

The prepolymer usually has a molecular weight of about 400−4000 and is further reacted with diamines or diols to extend the chain length. The final polymers can have excellent mechanical properties depending on the bulk structures. Hydrophilic segmented polyurethanes are synthesized using polyether diols, such as poly(ethylene oxide), polypropylene oxide, and polytetramethylene oxide, as the soft segment. The hard segment is created from ethylene diamine, tolylene diisocyanate, or 4,4'-diphenylmethane diisocyanate.

Gould and Kliment [7] prepared a hydrophilic polyurethane as follows. To a reaction mixture comprised of CARBOWAX [poly(ethylene glycol)]-1450 (55.5%), diethylene glycol (8.8%), and water (0.3%), were added methylenebis(cyclohexyl-4,4'-isocyanate) (35.4%) and catalyst (stannous octoate) at 65°C under constant stirring, followed by curing at 100°C for one hour. The reaction produced a hydrophilic foamed polyurethane having a water content of 60% on hydration to equilibrium. A latex rubber tube was dip-coated with a solution mixture of the hydrophilic polyurethane and dichloromethane, and following the drying and curing, the coated tube was immersed in a sulfuric acid-glycerol solution for 4 seconds. After immediately washing the tube, the coefficient of friction was measured to be 0.01 to 0.02 by the ASTM D 1894 method.

Creasy et al. [8] showed that organic solvent-soluble thermoplastic polyurethane compositions could be blended or alloyed with PVP or other poly(N-vinyl lactams) by the use of a common solvent or by conventional melt blending techniques. For example, PVP was dissolved in a mixture of diacetone alcohol and methyl ethyl ketone. The substrates coated with the above solution did not lose lubricity, even under continuous contact with water for six months. They further disclosed an improved method for creating a hydrophilic polymer blend from a first polymer component, that is, an organic solvent-soluble thermoplastic poly(vinyl butyral) and a second polymer component, PVP. The blend demonstrated a slipperiness in aqueous environments that was useful in low-friction coatings for a wide variety of substrates, among other applications [9].

Hydrophilic polyurethanes were also described by Hudgin [10], who claimed that they had many applications due to their high mechanical strength, high resistance to chemical attack, and low toxicity, so as to be compatible with the acids and enzymes found in the human stomach. Vailancourt [11] applied such a hydrophilic polyurethane to intubation devices such as nasogastric tubes. A hydroxy-terminated hydrophilic polyurethane having an average molecular weight of 7500 was prepared by reacting polyethylene glycol with methylene bis(4-cyclohexylisocyanate) in the presence of stannous octoate catalyst. He insisted that the coating was strong and attained sufficient lubricity upon contact with water in less than 5 minutes. Teffenhart [12] also prepared a hydrophilic, thermoplastic polyurethane blended with glycol components such as ethylene glycol, diethylene glycol, and polyoxyethylene glycol.

Lambert [13,14] invented a process for coating comprised of two steps. First, a 0.05 to 40% solution of a compound having at least two unreacted isocyanate groups per molecule was applied onto a substrate surface, and subsequently 0.5 to 50% solution of poly(ethylene oxide) was again applied onto the surface after evaporation of the isocyanate solution. Finally, the poly(ethylene oxide) solution was evaporated, followed by curing at an elevated temperature, preferably in the presence of a catalyst for curing. They also carried out hydrophilic coating with PVP [15].

3.4.2.1.2 PVP and Derivatives

The hydrophilic lubricating materials used until now have generally been PVP-polyisocyanate interpolymers or their hydrogels [16]. However, PVP and related poly(N-vinyl)lactams are eventually leached out from the coating compositions when placed in contact with aqueous fluids.

Schwartz et al. have applied PVP on a poly(vinyl chloride) (PVC) substrate using dimethylformamide (DMF) as a solvent for PVP [17]. The PVC tubing plugged at one end was dipped into the DMF solution of PVP for 5 seconds at room temperature and dried by a hot air gun at 200°C for 1 min. The coefficient of friction of the coated tubing was found to be 0.1 after immersion in water for several seconds. A similar combination was disclosed by Johansson and Utas-Sjoeberg [18]. Their method for forming an improved hydrophilic coating is to apply a nonreactive hydrophilic polymer layer with an osmolality-increasing compound to produce a more slippery surface under a wet condition. The polymer surface layer is treated with a solution of an osmolality-increasing compound having a concentration above 2 wt%. The osmolality-increasing compounds include glucose, sorbitol, sodium chloride, sodium citrate, sodium benzoate, calcium chloride, and so on. The results are shown in Table 3.1. Kliment and Seems [19] carried out hydrophilic coating by a two-step procedure. Material surfaces having low coefficients of friction in the wet state were produced by applying adherent isocyanate coating onto the substrate, followed by a second coating of PVP copolymerized with a minor amount of an ethylenic monomer having active hydrogens, and curing the coated substrate to allow a reaction to proceed between the active hydrogens and the isocyanate groups.

TABLE 3.1. Catheters Coated with an
Osmolality-Increasing Compound [18].

PVP-Coated Catheter Treated with the Following Solutions	Slipperiness after Drying				
	0 min	1	2	3	4
Untreated	8	7	5	2	1
20% NaCl + 5% PVP	8	8	8	8	7
15% KCl + 5% PVP	8	8	7	7	6
15% Na Citrate + 1% Keltrol®	8	8	8	8	8
15% KI	8	8	7	4	2
15% Glucose + 5% PVP	8	8	8	7	6
30% Sorbitol + 1% Keltrol®	8	8	8	7	7

®Keltrol, a xanthan gum, is a registered trademark of Kelco Co.
Result: Table 3.1 shows that the catheters having a coating of a nontoxic, osmolality-increasing compound retain their slipperiness for a longer time than the corresponding untreated surfaces, i.e., the coated catheters dry more slowly. The osmolality-increasing compounds prevent the hydrophilic polymer surface from desiccating.

Thus, catheters treated with an osmolality-increasing compound such as sodium chloride according to the invention dry more slowly than corresponding untreated catheters. The sodium chloride treated catheter keeps its slipperiness for a much longer time period, which is very desirable.

Medical tests show that the catheters applied with a coating of an osmolality-increasing compound such as sodium chloride are superior both at the insertion and at the removal of the catheter in the urethra.

3.4.2.1.3 Other Coating Methods

Other examples of lubricious surface coatings are as follows. Judd and Talty [20] employed a surface lubrication technique with a tubular food casing. The purpose of their invention was to make it easy to stuff oversized sausages into regenerated cellulosic casings with the use of a coating agent made of water-soluble cellulosic ethers. The cellulose derivatives developed include carboxy methyl cellulose, hydroxy propyl cellulose, and methyl or ethyl cellulose, in addition to antifriction adjuvants such as polyoxyethylene sorbitan ester of higher fatty acids ("Tween 80").

It is often difficult to keep the hydrophilic material coated onto a hydrophobic substrate surface without delamination for a long-term duration because of the weak adhesion at the coated interface. Another disadvantage of this technique is formation of cracks in the coating. Therefore, techniques for greater adhesion between the lubricious coating and the substrate have been developed, which include hydrophilic polymer blending, copolymerization, and interpenetrating polymer networks. Shelanski et al. [21] have made copolymers of PVP and various polyisocyanates which could retain many of the desirable properties of PVP while reducing the water solubility and increasing adhesion of the copolymers to wood, glass, metal, and others. The method of coating a polymeric substrate involves contacting it with a solution of PVP in a solvent selected from dimethyl formamide, butanone, methanol, tetrahydrofuran, and dimethyl acetamide, and evaporating the solvent from the substrate so that the surface of the substrate retains the coated PVP until the surface is no longer sticky.

Micklus et al. [22,23] coated a substrate with PVP-polyurethane interpolymer by applying polyisocyanate and polyurethane in a solvent such as methyl ethyl ketone (MEK) to the substrate, followed by evaporation of the solvent. If the substrate was polyurethane, coating was performed using a solution of isocyanate containing prepolymer and polyurethane with a solution of PVP. The isocyanate could be modified with chain extenders like diols to effectively produce a high molecular weight polyurethane *in situ*. The requirement that reactive isocyanates should be present introduced unavoidable chemical instability (pot life) and eliminated the possibility of utilizing aqueous or alcoholic solvents for preparation. Reactive isocyanates can deactivate many additives such as pharmaceuticals, surfactants, and dyes. Winn [24] also used similar compositions, but his method required a chemical reaction forming a covalent bond between isocyanates and active hydrogen groups reactive toward isocyanate on vinyl lactam or ethylene oxide polymers. A polyisocyanate coupling agent was applied to the surface from a solvent solution followed by applying the hydrophilic copolymer from a solvent solution. In certain instances, the copolymer and isocyanate could be simultaneously applied from a solvent solution. The resulting hydrophilic coating was thromboresistant, biocompatible and stable. A United Kingdom patent [25] describes a process involving more than ten steps for applying a hydrophilic coating, consisting of an interpolymer of PVP and polyurethane. This method seems to be too complicated to be suitable for large-scale production because each step took six hours to carry through and, moreover, cracks were often formed in the coated materials.

Becker et al. [26] prepared a water-soluble polyurethane by extending a polyurethane prepolymer having isocyanate end blocks with another polyether, and further extending it stepwise with poly(ethylene glycol), 1,4-butane diol, and ethylene diamine. Following the process, PVP was added to produce a thoroughly dispersed solution. The coated materials became slippery within five seconds after immersion in water, but maintained their lubricity only about one hour when immersed in water. Halpern and Gould [27,28] reported hydrophilic coating of plastics, particularly with an aqueous solution of a polysaccharide which flowed uniformly over the surface of an anchor film applied to the plastics. Polysaccharides from the group of hyaluronic acid and its salts, chondroitin sulfate, and agarose were employed and albumin was added to the aqueous solution to provide uniform wetting over the anchor film on the plastics.

For the friction-reduction coating, Schaper [29] used other synthetic polymers including ionene (ionic amine) polymers, which generally consist of repeating units represented by the formula:

$$-[-\overset{\overset{\textstyle R}{\textstyle |}}{\underset{\underset{\textstyle R'}{\textstyle |}}{N^+}} - R'' -]-\quad X^-$$

He described that an effective coating depended upon the basic cationic nature of the ionene polymer or upon the chemical and physical properties of the functional groups employed in preparing the functional ionene polymer compositions. Novel functional ionene compositions have been found to be useful in a number of different application areas, for instance, to modify the rheological properties of fluids used as friction reducers or turbulence suppressors.

Lubricious surfaces are greatly influenced by the water content in the interpolymer or coating materials. Thus, many efforts have been made to prepare a surface or a material that can absorb large quantities of water molecules. Hydrocolloidal dispersions of random interpolymer compositions have a capacity for absorbing water in amounts from 10 to 100 times their own weight. For example, Beede et al. [30,31] employed a coating consisting of a mixture of 2-hydroxy-3-methacryloyloxypropyl-trimethylammonium chloride, acrylic acid, acrylamide, and N,N'-methylenebis-acrylamide (cross-linking reagent) in aqueous solution. It was first purged with nitrogen gas and heated to 55°C, followed by the addition of a solution of ammonium persulfate. When the solution became viscous, methanol was added to the reaction mixture, which was then cooled.

3.4.2.2 IRRADIATION METHODS

3.4.2.2.1 Gamma Radiation

Utilizations of gamma radiation are aimed primarily at formation of hydrophilic hydrogels on polymer substrates, except for a few studies [32]. Hyans [33] disclosed a method for forming a self-lubricating fill tube having a hydrophilic polymer coating. The method consists of irradiating the fill tube with gamma radiation to provide the surface with radicals sufficient for initiating polymerization of monomer. Following irradiation, the tube was immersed in a bath containing a monomer and then heated to initiate polymerization. Fydelor et al. [34,35] prepared a hydrophilic, water-swell-able graft copolymer on an ethylene-vinyl acetate copolymer (EVA) (VA content 8 – 30 wt%) by radiation graft copolymerization of acrylic acid, and subsequently heating it in an aqueous hydroxide solution at temperatures higher than the softening point of EVA. They pointed out that, for surface graft copolymerization, the reaction should be performed in the absence of any solvent, i.e., in an undiluted acid. Surface grafting with the use of gamma radiation was also applied to a plastic tube having an internal longitudinal channel and holes extending from the channel to the outer surface of the tube to develop a surgical device for facilitating access to a bodily cavity [36]. The tube was normally fitted with a sleeve of a water-swellable material, especially a hydrophilic graft copolymer.

Ionizing and nonionizing radiation have also been utilized for coating with PVP. For instance, Merrill [37] disclosed a process whereby a silicone catheter was modified to have a hydrophilic surface by contacting it with N-vinyl pyrrolidone (NVP) monomer in bulk or in solution, followed by exposure to ionizing radiation at high dose rates. Penetration of NVP beyond a thin surface layer was prevented by controlling the dosage of the ionizing radiation and the concentration of NVP.

3.4.2.2.2 UV Light

Watts and Errede [38] disclosed a method for altering a hydrophobic film surface by coating it with a very thin layer of modifying material, bonded to the substrate by irradiation with UV light. A modifying material which gives the desired surface properties was present as an emulsion in a solution of a second modifier which was readily bondable to the substrate by UV light. In a typical example, 1% solution of PVP with added saponin was coated to a sheet of PET as a thin layer. The wet sheet was then heated to about 100°C and irradiated for 1 min with a germicidal ultraviolet lamp. A UV irradiation process was also applied to PEG cross-linking [39]. The coating material consisted of PEG, a cross-linking agent, a swelling solvent, and a radical initiator. For example, monoethyl-etherified PEG with a molecular weight of 550 was monoacrylated with acrylic acid. After evaporation to obtain the designated product, hexamethylene diol diacrylate (cross-linking agent), 2-hydroxy-2-propio-phenone (photoinitiator) and a solvent mixture consisting of ethanol, toluene, and tetrahydrofuran were added to the product. Each of the solutions was dropped on a PVC plate, allowed to evaporate, and then cured by a UV lamp.

Howard [40] pointed out that the use of cross-linked polymer was necessary in order to prepare solid-shaped objects having both long-term durability and lubricious properties under wet conditions. For this purpose, a solid surface was brought into contact with a solution containing at least 0.1 wt% of uncross-linked hydrophilic polymer and a free radical initiator selected from peroxides and photoinitiators. After air drying the coated surface, the dried coated surface was heated at the decomposition temperature of the free radical initiator, or exposed to UV in the presence of photoinitiator. Cross-linking was also effected by electron beam radiation or corona discharge.

3.5 REFERENCES

1. Ikada, Y., M. Suzuki and Y. Tamada. 1984. "Polymer Surfaces Possessing Minimal Interaction with Blood Components" in *Polymers as Biomaterials*, S. W. Shalaby, A. S. Hoffman, B. D. Ratner and T. A. Horbett, eds., New York: Plenum, pp. 135–147.
2. Thomas, J. J. and M. Sobel (Johnson & Johnson). U.S. patent 4381008, Apr. 26, 1983.
3. Seltzer, R. 1985. "Self-Adhesive Polymeric Coatings have Nonstick Surfaces," *C&EN* (Oct. 14):44–47.
4. Fan, Y. L. 1990. "Hydrophilic Lubricious Coatings for Medical Applications," *Amer. Chem. Soc., Polym. Mater. Sci. Eng.*, 63:709–716.
5. Wong, E. W. and D. G. Ballan. U.S. patent 4838876, June 13, 1989.
6. Althans, W. and J. Throne. EP 321679, June 28, 1989.
7. Gould, F. E. and C. K. Kliment. U.S. patent 4810543, Mar. 7, 1989.
8. Creasy, W. S., K. H. Lorenz and R. G. LaCasse. U.S. patent 4642267, Feb. 10, 1987.
9. Creasy, W. S. U.S. patent 4847324, July 11, 1989.
10. Hudgin, D. E. U.S. patent 3975350, Aug. 17, 1976.
11. Vailancourt, V. L. U.S. patent 4705511, Nov. 10, 1987.

12. Teffenhart, J. M. U.S. patent 4789720, Dec. 6, 1988.

13. Lambert, H. R. U.S. patent 4487808, Dec. 11, 1984.

14. Lambert, H. R. U.S. patent 4585666, Apr. 29, 1984.

15. Lambert, H. R. U.S. patent 4666437, May 19, 1987.

16. Riley, R. L., C. R. Lyons and U. Merten. 1970. "Transport Properties of Polyvinyl-pyrrolidone-Polyisocyanate Interpolymer Membranes," *Desalination*, 8:177–193.

17. Schwartz, A., J. Graper and J. Williams. U.S. patent 4589873, May 20, 1986.

18. Johansson, E. G. and J. M. R. Utas-Sjoeberg. EP 217771, Apr. 8, 1987.

19. Kliment, C. K. and G. E. Seems. U.S. patent 4729914, Mar. 8, 1988.

20. Judd, E. and R. K. Talty. U.S. patent 4169163, Sep. 25, 1979.

21. Shelanski, M. V. G. Mills and T. L. Wyndmoor. U.S. patent 3216983, Nov. 9, 1965.

22. Micklus, M. J. and D. T. Ou-Yang. U.S. patent 4100309, July 11, 1978.

23. Micklus, M. J. and D. T. Ou-Yang. U.S. patent 4119094, Oct. 10, 1978.

24. Winn, R. A. U.S. patent 4373009, Feb. 8, 1983.

25. Winn, R. A. U.K. patent 1600963, 1983.

26. Becker, L. F., D. G. Laurin and J. A. Palomo. U.S. patent 4835003, May 30, 1989.

27. Halpern, G. and J. U. Gould. U.S. patent 4657820, Apr. 14, 1987.

28. Halpern, G. and J. U. Gould. U.S. patent 4722867, Feb. 2, 1988.

29. Schaper, R. J. U.S. patent 4166894, Sept. 4, 1979.

30. Beede, C. H., H. L. Waldman and T. Blumig. U.S. patent 4248685, Feb. 3, 1981.

31. Beede, C. H., H. L. Waldman and T. Blumig. U.S. patent 4111922, Sep. 5, 1978.

32. Rosiak, J., 1983. *Radiat. Phys. Chem.*, 22:917; Rosiak, J., J. Olejniczak and A. Charlesby. 1987. "Determination of the Radiation Yield of Hydrogels Crosslinking," *Radiat. Phys. Chem.*, 32(5):691–694.

33. Hyans, T. E. U.S. patent 4459318, July 10, 1984.

34. Fydelor, P. J., R. A. Miller, B. J. Ringrose and W. A. Ramsay. U.S. patent 4785059, Nov. 15, 1988.

35. Fydelor, P. J., R. A. Miller, B. J. Ringrose, and W. A. Ramsay. EP 179839, May 7, 1986.

36. Miller, R. A. U.K. patent 2179258, Mar. 4, 1987.

37. Merrill, E. W. U.S. patent 4055682, Oct. 25, 1977.

38. Watts, R. E. and L. A. Errede. U.S. patent 3892575, July 1, 1975.

39. Goelauder, C., E. Joensson and T. Vladkove. EP 229066, July 22, 1987.

40. Howard, E. G. W.O. patent 8909246, Oct. 5, 1989.

CHAPTER 4

Lubricated Surfaces for Medical Use

As has already been mentioned in Chapter 3, biomaterials to be used for catheterization in urinary, tracheal, and cardiovascular tracts, or for endoscopy, should have a surface that has good handling characteristics when dry, but which preferably becomes slippery upon contact with aqueous body liquids. Such a low-friction surface would enable easy insertion and removal from a patient. It would further prevent mechanical injury to the mucous membranes and would minimize discomfort to the patient. In addition, particularly for artificial joints, durability against wear is of major importance and is related to the surface lubricity. Traditionally, lubricants and jelly-like materials have been applied onto biomaterial surfaces. Lidocaine jelly has been commonly used because of its additional local anesthetic effect. However, these substances cannot maintain a high degree of slipperiness for the required duration of time, due to leaching or dispersion into the surrounding body fluid. Therefore, a number of attempts have been made to render the polymer surfaces frictionless for long-term use, not only for catheters but also for other medical devices such as ophthalmologic materials. The idea of using hydrophilic polymer gels for contact lenses was proposed for the first time by Wichterle and Lim [1]. However, poly(2-hydroxyethyl methacrylate) (PHEMA) which was invented and used as a soft contact lens by Wichterle and his group, displays no highly lubricious surface even when it is hydrated, as the water content of this hydrogel is less than 40 wt%. Nevertheless, the PHEMA hydrogel has a better biocompatibility than rigid poly(methyl methacrylate) (PMMA). The term ''biocompatibility'' involves a wide variety of properties, as represented in Figure 4.1. Among the more well-known properties are bioinertness, antithrombogenicity (thromboresistance), and bioadhesion towards hard and soft tissues. Although biocompatibility is very important and hence has attracted much attention, the present chapter does not deal with it, but rather describes the characteristics of lubricated surfaces as they relate to biomaterials.

When compared to the hydrophilic polymer coating methods, the surface grafting of water-soluble polymers is much more effective for obtaining a

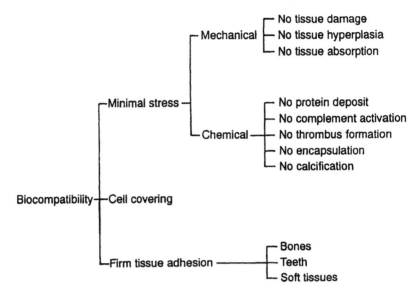

FIGURE 4.1. Classification of biocompatibility.

sufficiently lubricious surface when brought into contact with water. However, the process of grafting is somewhat complicated and rather expensive for commercial scale production. Until now, no detailed report has been available showing successful manufacturing of medical devices having a slippery surface obtained by surface graft polymerization. Hoffman [2] extensively studied modification of the surfaces of medical devices using ionizing radiation. For instance, he and Ratner [3] described the graft polymerization of NVP to polymer substrates in the presence of cupric or ferric ions after pretreatment of the substrate surface by ionizing radiation. However, the use of ionizing radiation is rather expensive and is not easily suited for large industrial scale production. Therefore, a large number of investigations have focused on attaining a hydrophilic lubricious surface through various coating procedures. Some of the more significant studies on lubricious polymer surfaces published in patents and academic journals associated with biomedical applications are presented below.

4.1 GUIDEWIRE

Heidrich stated at the 11th Cardiovascular Meeting in Lubek, Germany, in 1990 that the rate of occurrence of aneurysm is more frequent than expected when catheters or guidewires are placed into the blood vessels of patients. Specifically, more than 5% of the patients suffered from the disease, which occurred when balloon catheters were inserted and subsequently enlarged in the blood vessels or when they were used in the arteria femoralis. It was suggested that the main cause of this was the injuring of the cardiovascular endothelium, which had been brought about in the course of the insertion of the catheters or the guidewires.

Heilman and Waddell [4] presented an invention related to manufacturing guidewires having a wound outer casing with a smooth surface. One example of the guidewires was developed from a coiled semirectangular flatwire coated with a surface lubricant such as Teflon prior to winding. In another example, the flatwire was wound, polished by abrasion, and then electropolished. A combination safety and core wire, extended longitudinally within the coiled guidewire, was welded to the respective ends of the guidewire. The safety core wire was cylindrical and its uniform main body was smoothly tapered to a flexible, flattened, distal tip by means of combined mechanical metal-forming and electro-etching techniques (Figure 4.2).

Kikuchi et al. [5] described a guidewire constructed in one piece from a titanium-nickel alloy core and two adherent layers of plastics. The core was kink-resistant, and the inner coating layer was made from polyurethane

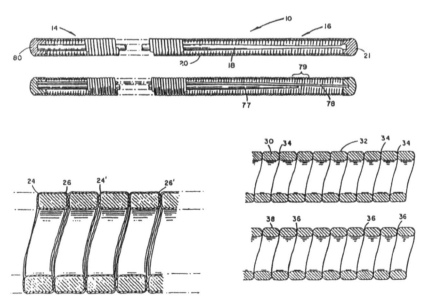

FIGURE 4.2. Guidewire having a smooth surface. 10: guidewire; 14: a proximal portion; 16: a distal tip; 18: combined core and safety wire; 21, 80: ends; 77: core wire; 78: safety wire; 79: part which is subject to preferential flexion; 24, 24′, 26, 26′: edges of the flatwire; 32: surface irregularity; 34, 36: inter-coil space; 38: surface which is virtually free from irregularity.

impregnated with tungsten microparticles to provide good fluoroscopic visibility. The outer coating layer was made from a hydrophilic polymer, which resulted in a low-friction surface when wet. The surface of this guidewire was smooth because of its polyurethane covering, and the hydrophilic polymer coating over the polyurethane layer provided lower surface friction than the Teflon-coated conventional guidewires. The composition of the hydrophilic polymer used for the coating was not disclosed in the article.

Iwatschenko [6] described a hydrophilic guide that could be removed from a body cavity at a force of 0.5 to 1 N. The invention used gelatin as the coating material, with a combination of alcohol as a softening agent. Sorbite could be used as well as polyhydric alcohols such as glycerine and generally, an addition of 0.5 − 5% glycerine was sufficient. If necessary, an addition of formaldehyde or another aldehyde to the gelatin was recommended in order to reduce the solubility of the gelatin.

A lubricious guidewire coated with an organosiloxane copolymer was patented by Gold [7]. In the patent, the core wire was inserted into the outer casing and attached to the casing at the distal and proximal ends. The distal tip of the core wire was tapered to a very small cylindrical cross-sectional area in order to make the guidewire flexible at the distal tip. The outer surface of the wound guidewire was then coated with copolymers of methylsiloxane and aminoalkylsiloxane units.

Terumo Inc. prepared a lubricious surface guidewire by bonding hydrophilic polymers containing maleic anhydride to the surface of the wire using polar groups such as maleic anhydride polymer or copolymer consisting of other hydrophilic components [8]. According to this technique, the substrate surface should be provided at least with a polar group such as aldehyde, epoxy, isocyanate, or an amino group. Provided the polymeric coating of the guidewire contained no such polar groups, it could be introduced just prior to the chemical reaction with maleic anhydride derivatives. The coefficient of static friction of the modified surface of this guidewire is around 0.02 when measured by a sliding angle method.

4.2 CATHETER

Butler and Kunin [9] pointed out the importance of prevention of periurethral contamination during catheterization, as well as the importance of surface lubricity of the catheter. They evaluated catheters impregnated with a jelly-like lubricant and antimicrobial agents such as polymyxin B and benzalkonium chloride.

Cohen [10] also emphasized the importance of the surface of the catheter to prevent microbial contamination of the bladder. He performed a catheterization test on thirty normal adult volunteers using catheters treated with povidone-iodine lubrication gel (PVP-I, "Betadine" lubrication gel), and showed the efficacy of PVP-I. The mean colony counts per 5 ml urine after the passage of the catheter coated with PVP-I was 3.6 compared to 65.61 for catheters that were coated with a normal jelly (K-Y brand lubricating jelly from Johnson & Johnson). Nevertheless, the mechanical friction between the catheter and the mucosa may injure both male and female urethrae. Bleeding is often reported, but in many cases, the injury is not immediately observable. Friction can further cause microhematuria, which, in turn, may lead to infection, stenosis, or some other complicated problems. Therefore, many studies have been directed to preparing a hydrophilic lubricious surface for catheter materials.

A low-frictional urinary catheter commercially known as LoFric® has been developed and manufactured by Astra Meditec Inc., Sweden. It has a hydrophilic layer consisting of PVP which enables the surface to bind water. The catheter tubing is made of medical PVC, which is nontoxic and pyrogen-free, and sterilization is performed by ethylene oxide gas. The catheter tip and eyes are slightly rounded to reduce the risk of trauma. The structure of this catheter is schematically depicted in Figure 4.3. The LoFric catheter is dipped in sterile water, physiological saline, or a chlorhexidine solution for 30 seconds just before use. A clinical study involving 671 catheterizations, which were mostly for male patients (and in some cases, performed by themselves) revealed that more than 76% of the patients preferred the LoFric catheter than the conventional ones.

To develop low-frictional catheters, tubes of an ethylene-vinyl acetate copolymer (EVA) and a plasticized PVC were also modified by photo-induced graft polymerization of *N*,*N*-dimethyl acrylamide (DMAA) [11]. In contrast to the plasticized PVC, EVA provides a good thermoplastic elastomer without using any additives such as plasticizer and, hence, has been employed concurrently for blood bag and drug delivery system in the medical field. According to the technique, the EVA or PVC tubes were first UV preirradiated in air with a germicidal or low-pressure mercury lamp for 30 min under continuous rotation, in order to be irradiated homogeneously.

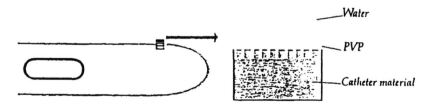

FIGURE 4.3. The structure of the LoFric® catheter.

The irradiated tubes were again UV irradiated with a high-pressure mercury lamp to effect graft polymerization in a Pyrex glass tube in which the catheter tubes were immersed in an aqueous monomer solution containing riboflavin. In the presence of riboflavin, surface graft polymerization onto the substrate tubes was successfully performed without any degassing procedure, because riboflavin consumed the oxygen molecules present in the monomer aqueous solution upon UV irradiation [12]. The advantage of this technique includes hydrophilic lubricious modification that is permanent, a process that can be used on a limited surface region without causing deterioration of the bulk properties and avoiding the time-consuming degassing procedure necessary for other grafting techniques based on radical polymerization. The cross-section of a grafted EVA tube is shown in Figure 4 .4. It should be noted that the LD_{50} of DMAA monomer is almost twice or triple that of acrylamide [13].

A flexible introducer sheath tube for including catheters was prepared by Kocak using a PVP-polyurethane blend to impart lubricity and thromboresistance [14]. The coating composition consisted of a thermoplastic polymeric material dissolved in a solvent. The sheath was dipped in the solution and then dried.

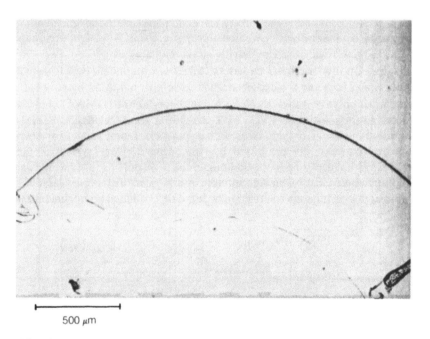

500 μm

FIGURE 4.4. Optical micrograph of the cross-section of ethylene-vinyl acetate copolymer tube grafted with *N,N* dimethylacrylamide and subsequently stained with rhodamin B.

Stoy et al. [15] prepared surgical tubular devices such as catheters consisting of a hydrophilic copolymer of acrylonitrile with either acrylamide, acrylic acid, or other comonomers. The copolymers were swellable in water. In the swelled condition they were pliable, elastic, and strong. Their properties could be altered by changing the degree of hydrolysis of polyacrylonitrile, or by changing the content of hydrophilic units. The part expected to be exposed to the atmosphere during the application to the patient was permanently protected against drying by a layer of polymer or copolymer impermeable to water and water vapor.

Cox [16] examined the hydrated surface of a hydrogel-coated latex urinary catheter using low-temperature SEM, and compared its appearance with that of some other catheter surfaces. They found that the effect of coating was to smooth over most of the ripples and fissures of the underlying latex. On hydration, the hydrogel surface became even smoother, with shallower furrows. The surface topography compared favorably with that of pure silicone, which has a uniformly rippled, but smooth appearance. On the other hand, Haindl and Haacke [17] pointed out that the chemistry of a latex surface is more important than the hydrogel-coated geometry for lower friction and that a hydrophilic coating does not change the surface properties of the underlying materials.

4.3 ARTIFICIAL JOINT

In 1966, McKee and Watson-Farrar [18] developed an artificial hip joint consisting of a metal acetabular cup and a metal femoral head. Although patients could stand and walk, other problems such as wear and corrosion friction took place, which remain unsolved today. Metal alloys and alumina ceramics have been used as the artificial femoral head, but the problems arising from the wear and friction have not yet been resolved. It should be noted, however, that the friction in natural human joints takes place between the soft articular cartilage in the presence of the joint fluid (synovia) containing proteoglycans such as hyaluronic acid. Thus no solid-solid direct contact occurs in the natural system between acetabulum and femoral head. The surprisingly low friction in natural joints may be explained in terms of fluid-film lubrication in which a minimal contact occurs between the joint surfaces, and almost all the load is carried by the fluid film. As is widely accepted, in the fluid-film lubrication, the fluid film maintains its effect as long as energy is continuously supplied for the liquid film to exist between the two solid surfaces. Otherwise, this liquid film layer will be lost gradually because of squeeze film effect under the everlasting compression.

As can be seen in Figure 4.5, currently used artificial hip joints with bone cement involve five interfaces with a marked difference in mechanical properties between the directly contacting surfaces. They are bone to bone cement, bone cement to PE, PE to metal, metal to bone cement, bone cement to bone. In fact, these five interfaces will become a cause of loosening, because each is bonded mechanically by different materials.

In 1982, Stauffer [19] reported that 26 (11.6%) of the acetabular and 69 (29.9%) of the femoral components of 231 Charnley prostheses showed radiographic evidence for loosening at ten years postoperation. Since most of the loosening occurred at the interface between bone and bone cement, the loosening was attributed mainly to the use of bone cement. Thus, many attempts have been made to develop cementless artificial joints. However, such incidences of loosening have not yet been resolved to date.

Oka et al. [20] pointed out that although a roentgenogram taken soon after bilateral total hip-replacement revealed that operations were apparently carried out without major technical failure, the joint components examined nine years after the operation had undergone marked proximal migration due to prominent bone destruction around the joint. Therefore, an attempt was made to develop an artificial articular cartilage from a new viewpoint of joint biomechanics, in which lubrication and load-bearing mechanisms of natural and artificial joints were taken into consideration. For this purpose, they selected a PVA hydrogel that exhibits viscoelastic behavior like an elastomer. Since the lubrication of the hydrogel is related to the fluid

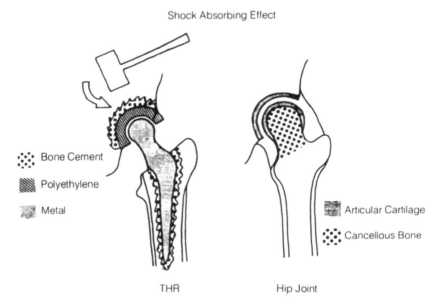

FIGURE 4.5. Shock absorbing effects of natural and artificial joints. (Source: Oka et al., 1990 [20].)

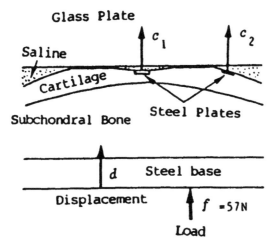

FIGURE 4.6. Cross-sectional view of the array of a glass plate and an osteochondral specimen. The specimen was pressed to the glass plate. Two thin steel plates (C_1, C_2) were attached to the cartilage surface as a target for the laser beam. (Source: Oka et al., 1990 [20].)

layer on its surface, Oka et al. measured the change of the thickness and the fluid pressure in the gap formed between a glass plate and the specimen under loading. It was found that the PVA hydrogel had a lower peak stress and a longer duration of sustained stress than PE, suggesting a better damping effect (Figure 4.6).

4.4 SUTURE

The suture is an essential element for surgical operations. Among its required properties is lubrication, which is needed for good handling, especially for smooth sliding-down of knots. Hunter and Thompson [21] measured surface performances of sutures such as roughness and sliding characteristics using an INSTRON testing machine. They recommended that sutures be coated with an aliphatic polyester such as a condensate of adipic acid and 1,4-butanediol having a molecular weight of about 2000 – 3000 for good "tie-down performance" (the ease of sliding a single throw knot into place down the suture). Collagen-based surgical threads, in the form of "catgut" sutures and ligatures, were employed as absorbable surgical threads in the late nineteenth century. Vivien and Schwartz [22] described collagen- or catgut-based sutures containing a significant amount of water through hygroscopic agents. The thread was treated with water and products capable of maintaining moisture on the surface of the thread. The products they used were fatty compounds and derivatives, such as glycerine, or polyoxyalkylenes such as polyethylene glycol. Coatings of hydrophobic agents were selected from lipids and silicones. Wilson and Simon [23] also disclosed a collagen-based thread containing water absorbed through the use of a hygroscopic agent and a nonionic surfactant in an amount sufficient to enhance its pliability, lubricity, and tie-down properties. Messier and Rhum [24] coated a braided multifilament type of surgical suture with a lubricant agent selected from a high molecular weight polycaprolactone or copolymers of other biodegradable monomers. The lubrication coating was made on the surgical filament consisting of a diblock copolymer having a first block of polyalkylene oxide and a second block of glycolic acid ester and trimethylene carbonate linkages. A triblock copolymer was also used that had a middle block obtained from an ethylene oxide homopolymer or a copolymer of ethylene oxide and a cyclic ether.

Tie-down properties of multifilament surgical sutures were improved by Shalaby and Jarniolkows [25] by coating the sutures with an absorbable composition consisting of poly(alkylene oxalate) where the alkylene was either C_6 or a mixture of C_4 to C_{12} groups. They reported that the braided sutures coated with 1 to 15 wt% of this composition had the characteristics of smooth knot tie-down under both wet and dry conditions.

Cresswell and Johnstone [26] studied a lubricated thread using a lubricating film from phosphatides such as lecithin. The preferred agent for the emulsification of the phosphatides in the spinning bath was N-octadecyl tetrasodium N(1,2-dicarboxy-ethyl) sulfosuccinamate, which has the following structure:

$$
\begin{array}{c}
\text{CH}_2-\text{COONa} \\
\diagdown \\
\text{SO}_3\text{Na} \quad \text{CH}-\text{COONa} \\
| \quad \diagup \\
\text{HCCON} \\
| \quad \diagdown \\
\quad \quad \text{C}_{18}\text{H}_{37} \\
| \\
\text{CH}_2-\text{COONa}
\end{array}
$$

Such succinamate was particularly suitable since the large concentration of salts present in the spinning bath did not appear to detract from its efficiency in emulsifying the lubricant. Nichols [27] proposed the coating of silk sutures with poly(alkyl methacrylate). Glick [28] coated silk and other nonabsorbable synthetic filaments such as nylon with silicone resin. Perciaccante and Landi improved the knot run-down characteristics for synthetic bioabsorbable [29] and nonbioabsorbable surgical sutures [30] by coating them with a lubricating film of bioabsorbable copolymer having polyoxyethylene and polyoxypropylene blocks. This lubricant coating not only aided in the knot run-down characteristics but also increased the smoothness and flexibility of the sutures. Lubricant copolymers were absorbed from the suture within a few days. When the lubricant was absorbed in living tissue, the resistance of the knot to slippage or untying due to tissue movement was soon increased. Lehmann et al. [31] coated a suture and ligature with calcium stearoyl-2-lactylate or diblock and triblock copolymer having glycolic acid ester and trimethylene carbonate linkage. The general structure of stearoyl esters used was:

$$[C_{17}H_{35}COO(CHCO_2)_x]_2 - M$$
$$|$$
$$CH_3$$

where M represents an alkaline-earth metal. They proposed also to use diblock and triblock copolymers. The diblock copolymer consisted of polyalkylene oxide and glycolic acid ester, while the triblock copolymer had, in addition to these blocks, a middle block of ethylene oxide homopolymer or copolymer of ethylene oxide and cyclic ether [32]. Trubitsina et al. [33] used an emulsion of polyethylsiloxane and methylcellulose mixture as suture coating material. The mixture consisted of polyethylsiloxane having a molecular weight of $1200-1800$ ($1.5-3.0$ wt%), methylcellulose ($0.07-0.5$ wt%), and water.

4.5 CONTACT LENS

Hydrophilicity and lubricity are also important for the surface of contact lenses and intraocular lenses. Sulc and Krcova [34] hydrolyzed substrate surfaces with a strong acid at elevated temperatures to produce a thin hydrophilic surface. The substrates they used were homopolymers and copolymers based on acrylate, methacrylate, acrylonitrile and/or methacrylonitrile. According to their method, for instance, a contact lens of PMMA was immersed in a solution of sodium hydrosulfate with sulfuric acid (94%) at 110°C for 20 seconds, followed by immersion in a 5% potassium carbonate for 4 hours at 50°C, after washing it with water. Beavers [35] modified contact lenses with a hydrophilic coating. According to his invention, hard and soft contact lenses as well as intraocular lenses were provided with immobilized hydrophilic coating. An acrylic copolymer matrix in the form of a film was cured upon the clear lens base so as to allow it to react chemically with polysaccharide. Then the polysaccharide coating was cross-linked by chemical reaction with the copolymer film, so as to make it substantially insoluble and immobilized. As a result, the hydrophilic coating became highly lubricious and permanently immobilized on the matrix. Beavers and coworkers also obtained hydrophilic coatings using a mucopolysaccharide film [36]. A preferred method includes first coating the plastic base with an aqueous solution of mucopolysaccharide, drying by applying a water-miscible solvent, then cross-linking, and permanently immobilizing the first coating on the plastics by applying a solution of catalyzed organic soluble aliphatic polyisocyanate. Neefe developed an oxygen-permeable contact lens having a wettable surface [37]. An oxygen-permeable silane-methacrylate copolymer was obtained by copolymerization of methyl methacrylate and organosilane monomer such as Y-methacryloxypropyl trimethoxysilane having the structure:

$$H_2C = C-COO\ (CH_2)_3\ Si\ (OCH_3)_3$$
$$|$$
$$CH_3$$

Copolymerization was also performed for contact lenses with the combination of aryl ester, vinyl acetate, and maleic anhydride [38].

4.6 ORTHOPEDIC CASTING TAPE

Synthetic orthopedic casting tapes are produced using curable resins coated on a substrate such as fiberglass, polyester, or other polymeric materials. Until cured, these resins remain tacky, which makes it difficult to mold the cast to the patient's limb as the curing resin tends to stick to the protective glove worn by the cast applier. The reason for this difficulty is based on the fact that the glove sticks to the resin during the process of smoothing the cast with a gloved hand, as well as while holding the cast at certain points until it hardens.

Pike [39] used beeswax as a release agent in the formation of an immobilizing orthopedic structure reinforced with a methacrylate polymer. He also described the method of preparing a hard immobilizing orthopedic structure. Boardman [40] prepared an orthopedic bandage consisting of a flexible carrier supporting a water-soluble vinyl monomer selected from diacetone acrylamide, isopropyl diacrylamide, and their mixture. In preparing the bandage, the materials were dipped in water in the presence of a catalyst to initiate polymerization of the vinyl monomers and then the body portion was wrapped to be immobilized. In the preferred practice, the initiator was a part of the bandage and might either be mixed with the monomer or coated on the surface of the bandage. Sholz et al. [41]

developed orthopedic casting materials having reduced tack, which consisted of a nonwoven stretchable fabric impregnated with a curable resin, giving a kinetic coefficient of friction less than about 1.2 (when determined by ASTM D 1894 between test specimen and 200 g stainless steel sled). In order to reduce this friction even further, a lubricant is added to the materials which consists of hydrophilic groups covalently bondable to the curable resin. The lubricant to be added is preferably selected from any mixture consisting of a surfactant, a polymer having plural hydrophilic groups, and polysiloxane. Examples of monomers for such polymers having hydrophilic groups include acrylamide, vinylpyrrolidone, vinylacetate and its polymeric hydrolyzed derivatives, hydroxy and amino functional lower alkyl acrylates such as 2-hydroxyethyl acrylate, and various specialty monomers containing ionic species.

A polyurethane casting tape having controlled tackiness provided by incorporation of a water-soluble polymer into the polyurethane prepolymer of the casting tape was developed by Yoon [42]. The water-soluble polymer had PEG chains.

4.7 SURGICAL GLOVE POWDER

Natural and synthetic rubbers used for surgical products such as tubing, catheters, drains, and gloves are subject to sticking during storage. To prevent such sticking, various powders are commonly used as solid-film lubricants. The powders applied on surgical gloves should be nontoxic to living tissues and easily sterilized before use. It is further desired that these powders be biodegradable. Among the surgical glove powders, talc was most commonly used in the earlier application. However, the use of talc was soon replaced by starch because talc was found to cause granulomas in the body. Although starch has been pointed out to have a number of disadvantages, it is still being used as a solid-film lubricant. Numerous substitutions for starch were proposed as lubricating fine powder, including poly(glycolic acid) [43,44], chitin and its derivatives [45], poly(N-acetyl-D-glucosamine) [46], and an enzymatically degradable form of poly(N-acetyl-D-glucosamine). These powders are readily absorbed into living tissues, thus minimizing tissue reactions that may occur when the powders transfer from the surgical products such as the glove to internal sites [47].

Podell and Podell [48] developed a rubber glove that was laminated with an internal plastic lining of a hydrophilic material for reducing the friction between the glove and the hand of the user. They also developed a process consisting of immersing the elastomer in a concentrated solution of a strong acid such as sulfuric acid, washing, and subsequently dipping the treated elastomer in a solution of an uncured hydrophilic polymer. The treated

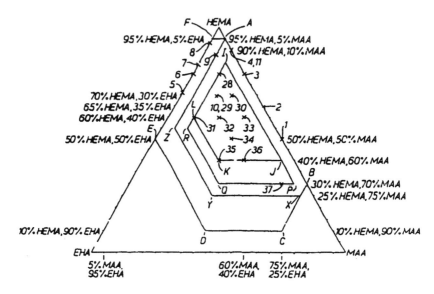

FIGURE 4.7. Diagram of components for coating solution used to treat gloves with low friction. (Source: Podell [50].)

elastomer was then held at an elevated temperature to cure the polymer coating and bond it to the elastomer. They also invented similar coating methods using coating solutions containing copolymer produced from a mixture of 80 wt % HEMA, 20 wt % 2-ethylhexyl acrylate, a catalytic curing agent of dicyclopentadiene diepoxide, and *para*-toluene sulfonic acid monohydrate. The concentration of the above coating mixture is 1−4 wt % in organic solvent such as methoxyethanol and ethanol [49]. They found also that some HEMA copolymers were superior in lubricity against dry skin. Examples of such hydrogel copolymers are HEMA and methacrylic acid (MAA) or 2-ethylhexyl acrylate (EHA), or a ternary copolymer of HEMA, MAA, and EHA [50]. A condensed diagram showing possible compositions of these three components is shown in Figure 4.7. The area ABCDEF covers substantially all of the hydrogel-forming area, apart from HEMA homopolymer and HEMA copolymers with up to 5 % EHA and/or MAA.

4.8 VIAL STOPPER AND OTHERS

For many years, the most successful closure system for pharmaceutical products consisted of using rubber stoppers in glass or high-density plastic vials. When liquids are contained in a vial, a needle can easily penetrate the rubber stopper to withdraw the desired amount of ingredient without otherwise interfering with the completeness of the closure. Romberg [51] made a convenient stopper device coated with a polyurethane film in order to improve the coefficient of friction below 0.6.

Hanke [52] proposed a lubrication compound suitable for hygienic and medical applications, such as for coating tampons and suppository structures. The compound contained a thermoplastic film-forming, water-soluble polymer, a plasticizer compatible with the polymer, and a lubricant. A preferred thermoplastic polymer was hydroxy propyl cellulose, and preferred plasticizers were olefinic glycols such as PEG and poly(propylene glycol).

Brook [53] prepared hydrophilic copolymers for wound dressings as well as for coating catheters. According to his invention, the hydrophilic product for use in contact with body tissue is comprised of a layer of an essentially uncross-linked copolymer and hydrophobic and hydrophilic monomer components. For example, the hydrophobic monomers were esters of methacrylic and acrylic acid, and the preferred hydrophilic monomers were vinyl pyrrolidone (VP) and HEMA. The advantage of this wound dressing was that the hydrogel layer was generally transparent so that the progress of the wound healing could be monitored without removing the dressing. Water uptake of the dressing was as high as 590% of the dry weight of the copolymer material. A highly flexible nature enabled it to conform to the wound area.

4.9 REFERENCES

1. Wichterle, O. and D. Lim. 1960. *Nature*, 185:117.

2. Hoffman, A. S. 1983. *Radiat. Phys. Chem.*, 22:917.

3. Ratner, B. D. and A. S. Hoffman. 1976. "Hydrogels for Medical and Related Applications," *Am. Chem. Soc. Symp. Ser.*, 31:1−36.

4. Heilman, M. S. and S. W. Waddell. U.S. patent 4080706, Mar. 28, 1978.

5. Kikuchi, Y., V. B. Graves, C. M. Strother, J. C. McDermott, S. G. Babel and A. B. Crummy. 1989. "A New Guidewire with Kink-Resistant Core and Low-Friction Coating," *Cardiovas. Intervent. Radiol.*, 12:107−109.

6. Iwatschenko. P. U.S. patent 4798593, Jan. 17, 1989.

7. Gold, P. U.S. patent 4534363, Aug. 13, 1985.

8. Terumo Inc. Japan patent 61-45775, March 5, 1986; Terumo Inc. Japan patent 1-33181, July 12, 1989.

9. Butler, H. K. and C. M. Kunin. 1965. "Evaluation of Polymyxin Catheter Lubricant and Impregnated Catheters," *J. Urol.*, 100(4):560−566.

10. Cohen, S. 1985. "A Microbiological Comparison of a Povidone-Iodine Lubricating Gel and a Control as Catheter Lubricants," *J. Hospital Infection*, 6:155−161.

11. Uyama, Y., H. Tadokoro and Y. Ikada. 1991. "Low-Frictional Catheter Materials by Photo-Induced Graft Polymerization," *Biomaterials*, 12:71−75.

12. Holmstrom, B. and G. Oster. 1961. "Riboflavin as an Electron Donor in Photochemical Reactions," *J. Am. Chem. Soc.*, 83:1867−1871.

13. Hayashi, I., M. Kunimoto, M. Terada and T. Tomita. 1974. "Toxicologic Investigation of Dimethylacrylamide in Mice," *Eisei Kagaku*, 20:317−321.

14. Kocak, N. U.S. patent 4705511, Nov. 10, 1987.

15. Stoy, A., V. Stoy and J. Zima. U.S. patent 4026296, May 31, 1977.

16. Cox, A. J. 1987. "Effect of a Hydrogel Coating on the Surface Topography of Latex-Based Urinary Catheters: An SEM Study," *Biomaterials*, 8:500−503.

17. Haindl and Haacke. 1989. "Hydrogel Coating of Urinary Catheters," *Biomaterials*, 10:215.

18. McKee, G. K. and J. Watson-Farrar. 1966. *J. Bone Joint Surgery*, 48B:245.

19. Stauffer, R. N. 1982. *J. Bone and Joint Surgery*, 64A:983.

20. Oka, M., T. Noguchi, P. Kumar, K. Ikeuchi, T. Yamamuro, S.-H. Hyon and Y. Ikada. 1990. "Development of an Articular Cartilage," *Clinical Mater.*, 6:361−381.

21. Hunter, A. W. and D. R. Thompson. U.S. patent 3942532, March 9, 1976.

22. Vivien, D. and G. Schwartz. U.S. patent 3896814, July 29, 1975.

23. Wilson, H. A. and B. Simon. EP Patent 128043, Dec. 12, 1984.

24. Messier, K. A. and J. P. Rhum. U.S. patent 4624256, Nov. 25, 1986.

25. Shalaby, S. W. and D. Jarniolkows. U.S. patent 4105034, Aug. 8, 1978.

26. Cresswell, A. S. and E. P. Johnstone. U.S. patent 2576576, Nov. 27, 1951.

27. Nichols. U.S. patent 2734506, 1956.

28. Glick, A. U.S. patent 3187752, June 8, 1965.

29. Perciaccante, V. A. and H. P. Landi. U.S. patent 4047533, Sept. 13, 1977.

30. Landi, H. P. and V. A. Perciaccante. U.S. patent 4043344, Aug. 23, 1977.

31. Lehmann, L. T., D. W. Wang, L. Rosatl and P. K. Jarrett. EP patent 258749, Mar. 9, 1988.

32. Casey, D. J., P. K. Jarrett and L. Rosati. U.S. patent 4716203, Dec. 29, 1987.

33. Trubitsina, N. I., O. A. Novikova, N. N. Gavrilyuk, N. B. Tichomirova, T. V. Spolnitsikaya, V. P. Sergeev, A. I. Malchevskyi and B. A. Egorov. U.S.S.R. patent 1265226, Oct. 23, 1986.

34. Sulc, J. and Z. Krcova. German patent 3841380, June 22, 1989.

35. Beavers, E. M. U.S. patent 4663233, May 5, 1987.

36. Halpern, G., C. Campbell, E. M. Beavers and H. Y. Cheh. U.S. patent 4801475, Jan. 31, 1989.

37. Neefe, C. W. U.S. patent 4280759, July 28, 1981.

38. Kato, H. Japan patent 54110289, Aug. 29, 1979.

39. Pike, G. J. U.S. patent 3089486, May 14, 1963.

40. Boardman, F. U.S. patent 3630194, Dec. 28, 1971.

41. Sholz, M. T., D. C. Bartizal, K. E. Reed, W. K. Larson, T. C. Sandvig, R. S. Buckanin, D. A. Ersfeld and P. E. Hensen. European patent 290207, Nov. 9, 1988.

42. Yoon, H. K. and R. L. J. Sun. EP patent 266892, May 11, 1988.

43. Semp, B. A. U.S. patent 3892314, July 1, 1975.
44. Ramsey, W. B and D. F. DeLapp. U.S. patent 3846382, Nov. 5, 1974.
45. Balassa, E. S. U.S. patent 3532754, Oct. 6, 1972.
46. Capozza, R. C. U.S. patent 3988411, Oct. 26, 1975.
47. Casey, D. J. U.S. patent 4068757, Jan. 17, 1978.
48. Podell, D. L and H. I. Podell. U.S. patent 3813695, June 4, 1974.
49. Goldstein, A. and H. I. Podell. U.S. patent 4482577, Nov. 13, 1984.
50. Podell, H. I., A. Goldstein and D. C. Blackley. U.S. patent 4575476, Mar. 11, 1986.
51. Romberg, V. G. U.S. patent 4756974, July 12, 1988.
52. Hanke, K. E. U.S. patent 3756238, Sept. 4, 1973.
53. Brook, M. G. U.S. patent 4842597, June 27, 1989.

CHAPTER 5

General Methods for Surface Modification

In the previous chapters (1 − 4) a survey of the published literatures including patents was presented concerning the state of the art of lubricated polymer surfaces. However, in order to develop much better lubricated polymer surfaces than those currently available, it may be necessary to understand at least the rudiments of the phenomena related to polymer surfaces and current methods for polymer surface modification. This chapter will be devoted to this purpose, without confining itself only to lubricated polymer surfaces.

5.1 POLYMER SURFACE STRUCTURE

When dealing with surface properties and structures of a material, only the outermost surface is enough to be taken into consideration, strictly speaking. However, this is not the case for polymer surfaces, as the physical structure of the outermost polymer surface is generally not fixed but continuously changing due to microscopic Brownian motion of the polymer segments. As mentioned in the previous chapters, a remarkable characteristic of the polymer surface is its high segmental mobility even at room temperature, in contrast to the rigid surfaces of metals and ceramics. This possible motion of the polymer segments suggests that a polymer surface cannot be described in terms of a depthless, two-dimensional plane, but only as a region with some thickness. As a consequence, we can look at the polymer surface in a direction perpendicular to the plane, in addition to that parallel to the surface plane.

Figure 5.1 shows a classification of polymer surfaces in the direction parallel to the surface plane. Surfaces of copolymers and blended polymers are likely to be heterogeneous in their horizontal chemical composition. These polymers must also have a variation in chemical composition in the vertical direction, which is called a depth profile, as shown schematically in Figure 5.2.

Most of the surfaces of polymers used in industry are not hydrophilic but hydrophobic. Therefore, it is difficult to directly bond these nonpolar polymer surfaces with other substances like adhesives, printing inks, and paints, which generally consist of polar compounds. However, the polymer surfaces are not so extremely hydrophobic as to completely reject adhesion of other objectives. It should be noted that a transparent polymer surface becomes translucent when exposed to high humidity at temperatures below the dew point, because of deposition of minute water droplets on the surface. An exception is regenerated cellulose film, which has lots of polar groups on and in the surface region. Static electrification is also closely related to the surface properties of polymers. For all of these reasons, surface modification of polymers has been an important technology in polymer applications since the advent of polymer industries.

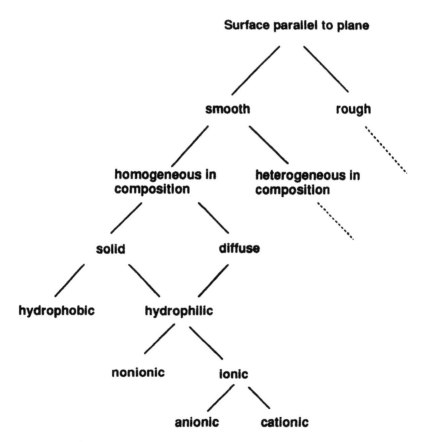

FIGURE 5.1. Classification of polymer surface contacting with water (parallel to the plane).

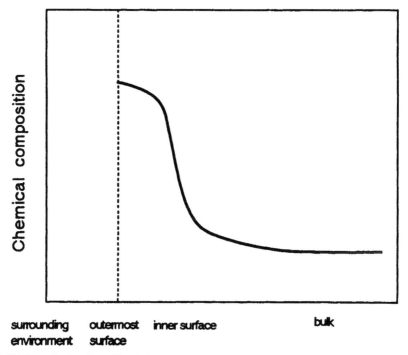

surrounding   outermost   inner surface   bulk
environment   surface

FIGURE 5.2. Depth profile of chemical composition of a polymer in the direction perpendicular to the surface plane.

5.2 THE PURPOSE OF POLYMER SURFACE MODIFICATIONS

The background and rationale for surface modifications of polymeric materials are completely different from those of metals. In the case of coating on metallic materials, surface treatment is mostly performed to protect the poor bulk properties (e.g., keeping iron from staining or erosion). The surface of iron can be covered by other metals that are less costly, have better mechanical properties, and simultaneously are more stable against staining. In other words, if the new surface resulting from the metal coating is satisfactory, the modified product can be used even though the bulk property is poor.

By contrast, surface modification for polymeric materials is quite different. In the case of polymers, their bulk properties should be excellent, regardless of surface modification. The purpose of surface modification for polymers is to impart new properties that the bulk polymer does not possess. The properties characteristic to polymer surfaces include wettability, adhesiveness, printability, electrostatic properties such as triboelectrification and electroconductibility, antifogging, antifouling, grazing, surface hardness, surface roughness, biocompatibility, lubricity, and so forth. Among these properties, wettability, printability, and adhesiveness can be generally improved by making the surface of polymers hydrophilic. Sumiya et al. showed that the adhesive force is almost proportional to the surface energy of the adherend [1]. The technologies and methods employed to improve the physical properties of polymer surface are briefly tabulated in Table 5.1, together with the nature of resultant surfaces. As can be seen, hydrophilicity of the surface is a very common property desired for many of these applications. Indeed, a large number of investigations have been devoted to rendering polymer surfaces hydrophilic.

As to surface lubricity, we have demonstrated that a more hydrophobic surface has a lower attractive force to the opposing surface, resulting in a smaller frictional force. Interestingly, a hydrated surface with larger amounts of water-soluble polymers immobilized on the surface region also interacts less strongly with the opposing surface [2]. Lubricious as well as antifogging and antifouling surfaces are not necessarily hydrophilic, as shown in Table 5.1. For instance, polytetrafluoroethylene (PTFE), one of

TABLE 5.1. Physical Properties of Polymer Surfaces and Methods for Their Modification.

Properties	Related Technology	Surface Nature and Related Items
Adhesion	polar groups anchoring	hydrophilic surface energy printability
Wettability	acidic, alkaline treatment	hydrophilic antifogging
Printability	corona treatment	hydrophilic adhesion
Antistaticity	metal-coweaving surfactant mixing	electroconductibility
Antifogging	hydrophilic-polymer coating heating extremely low surface energy	hydrophilic hydrophobic
Antifouling	hydrophilic-polymer coating micro-vibration painting with organic tin compounds extremely low surface energy	hydrophilic hydrophobic mechanical removal sea pollution
Biocompatibility	enzyme mixing protein immobilization surface grafting	hydrophilic hydrophobic bioinert antithrombogenicity
Lubricity	hydrophilic lubricious coating	hydrophilic hydrophobic

the most hydrophobic man-made polymeric materials with a water contact angle of around 120°, has a low coefficient of friction to make the surface less prone to fogging and fouling. However, as it is very difficult to find a much more hydrophobic surface than that of PTFE, almost all studies, with few exceptions, were directed to creating surfaces as hydrophilic as possible.

General surface modification methods can be classified into the following categories:

(1) Mixing polymers with low- or high-molecular weight compounds, accompanied by alteration of the bulk properties to some degree

(2) Controlling the physical structure, pore size, and microdomains in the surface region of the polymer, for instance, by the use of copolymers

(3) Coating with other materials having chemically and/or physically different properties

(4) Altering the surface composition of the polymer by formation or removal of functional groups on the polymer surface

Among the categories described above, (1) may not be considered as a surface modification, because the surface properties are almost identical to those of the bulk. Nevertheless, category (1) is important as a surface modification method, since surfactant mixing or hydrophilic polymer blending for antifogging, antistatic, and lubricious surfaces is also included in this category. As to category (2), we can mention as an example a urethane copolymer that contains a small amount of poly(dimethyl siloxane) (ca. 10%). Although it is claimed that this urethane copolymer is antithrombogenic when placed into contact with blood, it is often not so effective as claimed, probably because of the substantial absence of poly(dimethyl siloxane) moiety at the surface region in contact with an aqueous environment. Localization of polar or apolar groups in the surface region is known to depend on the nature of the surrounding environment [3].

Category (2) is not common. When a polymer is blended with microspherical particles, followed by subsequent removal of these particles by simple evaporation or extraction with an appropriate solvent, usually a pore-rich surface is obtained.

Categories (3) and (4) are the most important for surface modification of polymers, and they do not alter the bulk properties of the polymer. General methods for surface modification belonging to the categories (3) and (4) will be described below. We classify the polymer surface modification based on the particular modification technique used: either chemical, physical, or biological. The chemical modification is often called a wet process, while the physical modification is a dry process. Biological modification is aimed at imparting biological activities to the surface of man-made materials.

5.3 CHEMICAL MODIFICATION

5.3.1 Acidic and Alkaline Treatment

There are currently a variety of surface modification methods for polymeric materials. The most simple and primitive methods for rendering polymer surfaces hydrophilic are acidic and alkaline treatments, and flame exposure. They do not differ substantially from the modern dry methods such as treatment with corona and glow discharge or exposure to UV radiation, in terms of introducing oxidized hydrophilic polar groups onto polymer surfaces.

Treatment of an LDPE film with aqueous chromic acid solution was studied by Holmes-Farley et al. [4]. They presented evidence to support the existence of functional groups such as carboxylic acid, ketone, and aldehyde moieties in the "surface" and "subsurface" of the treated LDPE. Their conclusion was based on various analytical techniques including XPS, ATR-IR, titration, and others. The contact angle of films is also useful to study the effect of surface modification on wettability or hydrophilicity of the oxidized films.

Zeronian et al. [5] studied the hydrophilicity of PET film by measuring water contact angle. The hydrophilicity of the film decreased in the following order: PET treated with aqueous sodium hydroxide > untreated PET > PET treated with methanolic sodium methoxide. When the sodium-methoxide-treated polyester was hydrolyzed with caustic soda, its contact angle fell, indicating that the methyl ester groups formed during the sodium methoxide reaction by a base-catalyzed ester interchange had been the cause of the high contact angles. It appeared that carboxyl groups at the surface of hydrolyzed PET played a role in determining its hydrophilicity.

5.3.2 Direct Chemical Modification

When the polymer surface to be modified possesses reactive groups capable of combining other components such as water-soluble polymers, surface modification can be readily conducted by chemical reactions. Numerous synthetic reactions are available for this purpose. An example is the coupling of water-soluble polymers onto polymeric substrates having reactive groups like cellulose. PEG and its derivatives with terminal carboxyl or isocyanate groups can be covalently immobilized onto the surface of cellulose [6].

Complement activation induced by regenerated cellulose hemodialysis membranes was successfully reduced by surface modification with PEG [7]. This was accomplished by using an ester linkage obtained by the reaction of terminal carboxyl groups of the PEG chains and the hydroxyl groups on the membrane surface. In the modification method, PEG-diacid, dicyclohexylcarbodiimide (DCC), and dimethyl aminopyridine (DMAP) were dissolved in toluene. Cellulose films were immersed in the solution for 10 minutes, maintaining the temperature at 10°C, which was subsequently raised to 30°C to complete the reaction. PEG with different molecular weights (400, 1000, 4000, and 8000) were used. Among these polymers, the PEG with the 400 and 1000 molecular weights were found to give a good result for the decrease of complement activation. The hemodialyzer of cellulose hollow fibers produced with this modified surface exhibited good blood compatibility when used clinically for hemodialysis [8].

5.4 PHYSICAL MODIFICATIONS

The modification of polymer surfaces can be also achieved without using any wet chemical reagents – that is, through physical modification methods – where various high energy sources are employed. A variety of physical means used in this dry modification are summarized briefly in Figure 5.3.

5.4.1 Electron and Ion Beams

Electron beam irradiation of the high-modulus PE fibers was performed by Klein et al. [9]. They reported that the irradiation resulted in cross-linking in the amorphous regions and chain scission in the crystals. The gel content of the fiber irradiated in acetylene were much higher than for an equivalent dose in vacuum.

According to Technical News (*Biomedical Materials*, Elsevier, March 1991, p. 4), ion beam treatment to modify the surface properties of silicone rubber without changing its bulk properties has been attempted by Spire Corp. of Bedford, Massachusetts. To reduce the frictional properties of silicone rubber, they used a proprietary, low-temperature, dual ion beam-assisted process to deposit films ranging from 5 nm to several micrometers in thickness. The ion beam technology was used for surface modifications of metals, polymers, and ceramics.

5.4.2 Glow and Corona Discharge

"Plasma" means a mixed state of ionized gases that consists of electrons, ions, gas atoms, and molecules in either their ground or their excited state.

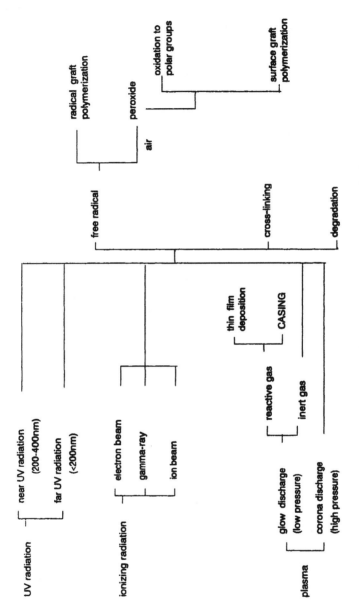

FIGURE 5.3. Physical methods used for surface modification of polymers.

The plasma state is generated by a high electric potential field and called "low-temperature plasma" if the electron temperature is not equilibrated with the surrounding environment. When the high electric energy is applied under a reduced pressure, the treatment is performed by the plasma generated by glow discharge, while the surface treatment may be a result of corona discharge if performed under the atmospheric pressure. Both the glow and corona discharges are currently used for surface modification of polymeric materials on an industrial scale. Corona discharge treatment may be far simpler than the glow discharge, because it does not require evacuation of the system.

Since glow discharge is performed under reduced pressure, not only is the so-called plasma treatment of substrate polymer surfaces possible, but so is the direct deposit of the organic layer onto the surfaces by introducing various gases. The formation of polymeric materials in the plasma environment is termed plasma polymerization. This polymerization reaction proceeds through a more complicated mechanism than the plasma treatment. In most cases, polymers are deposited in a thin film layer onto the substrate materials by plasma polymerization of monomer under glow discharge. The deposited polymers are highly cross-linked and branched. Yasuda et al. [10] attempted polymer deposition by plasma to modify the surface properties of contact lenses. A thin layer of plasma polymer from acetylene/H_2O/N_2 was deposited to PMMA contact lenses to a thickness of roughly 20 nm. Generally, plasma polymerization is not commonly utilized for surface modification to give hydrophilicity.

As is well known, plasma exposure alone usually makes hydrophobic surfaces more hydrophilic, although the acquired property does gradually disappear with time. If a polymer surface is exposed to plasma in the presence of ammonia gas, it is reported that amino groups are introduced onto the substrate surface. Hollahan and Stafford [11] exposed polypropylene, PVC, PTFE, polycarbonate, polyurethane, and PMMA to gaseous plasma of ammonia or nitrogen-hydrogen mixtures. In every case, the polymers became wettable after the treatment.

When conventional monomers such as vinyls and acryls are introduced to gaseous substances, plasma-induced polymerization—to be described below—takes place in addition to plasma polymerization. The polymer radicals formed upon plasma exposure are used to initiate radical polymerization when immediately brought into contact with monomers without exposure to air. This is called plasma-induced polymerization, and is often used for surface modification of polymers.

The plasma-induced polymerization can be initiated for surface graft polymerization of monomers not only through free radicals but also through peroxides. The radicals introduced onto the surface can react with atmospheric oxygen to yield peroxide groups, which are capable of generating peroxy radicals when heated or brought into contact with redox reagent. In 1961, Bamford and Ward [12] reported that free radicals could be produced on the surfaces of many polymer solids by subjecting them to a Tesla coil generating high-frequency electric discharge through a gas at low pressure. The radicals formed on the polymeric substrates could be used, either directly or indirectly, to initiate graft polymerization. They reported graft polymerization of acrylonitrile onto the surface of polyethylene and polypropylene.

Nuzzo and Smolinsky [13] described a procedure to modify the surface of PE film using a combination of plasma treatment and wet chemical technique, and evaluated the modified surfaces using contact angle and XPS measurements. The plasma treatment was carried out by exposing the PE film to oxygen, hydrogen, and water plasma. Water vapor or other gases were introduced into the reactor (Figure 5.4) to produce a total static pressure of 0.2 torr. The radio frequency of the generator supplied was 13.56 Mhz. For all the gasses, 5 W treatment for $1-2$ s produced approximately the same degree of surface modification. The XPS study showed that both oxygen and water plasma produced a variety of oxidation products ranging from alcohols to carboxylic acids. The wet chemical treatment was done using aqueous chromic acid prepared from chromium trioxide, water, and sulfuric acid in the ratio of 3:4:3 by weight. A PE film treated with oxygen plasma was floated on this solution for a constant duration of time at ambient temperature. This treatment oxidized the already plasma-oxidized surface further to give high densities of carboxylic acid groups.

More than ten kinds of polymeric materials were exposed to glow discharge, and the adhesive strengths of these treated materials were com-

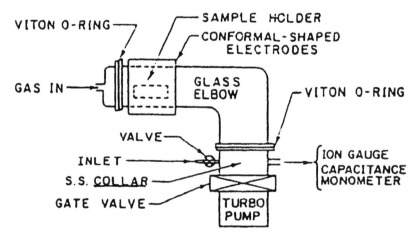

FIGURE 5.4. Schematic representation of the reaction chamber used by Nuzzo and Smolinsky [13].

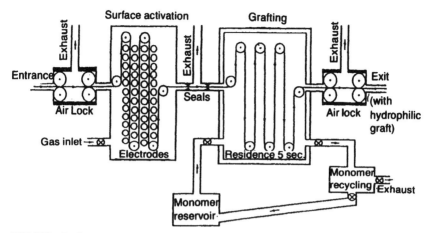

FIGURE 5.5. Proposed continuous grafting apparatus, plant scale; drive mechanisms and electrode connections not shown [15].

pared by Hall et al. [14]. Bondability of the films with an epoxy resin was measured with a tensile testing machine following exposure to helium or oxygen plasma. Generally, the bond strength increased rapidly with the exposure time and then remained nearly constant, or decreased in some cases at longer exposure times. Polypropylene showed a rapid increase in bondability after exposure to excited oxygen plasma. Helium was ineffective toward this polymer under normal exposure conditions, but could produce a good bond strength by plasma exposure at high temperatures.

Glow discharge has a disadvantage over corona discharge due to the reduced pressure when applied on a large industrial scale. In 1971, Bradley and Fales [15] proposed an apparatus as illustrated in Figure 5.5 for industrial applications of electrical discharge. This apparatus enabled graft polymerization onto plasma-treated surfaces on a continuous plant scale. However, to our knowledge no reports have appeared since then describing successful results of surface modification utilizing this technique. A most notable problem is probably the difficulty in evacuating whole systems and achieving a homogeneous treatment over the entire polymer surface at a low cost. Similar treatment by vacuum glow discharge was claimed by Tamaki and Tatsuta [16].

5.4.3 Corona Exposure

Typical apparati for corona discharge treatment are shown in Figures 5.6 and 5.7 for industrial and experimental purposes, respectively. Most corona discharge treatments have been performed under ambient atmospheric conditions. The exception can be found in the work of Steinhauser and

Ellinghorst [17]. They carried out corona discharge treatment on an isotactic polypropylene surface in nitrogen and carbon dioxide, and investigated the surface changes by contact angle and XPS measurements. It was shown that in the nitrogen discharge the polymer radicals could only react with atomic or excited nitrogen, whereas in CO_2, they also reacted with ground state molecules. When the samples were brought into contact with air after discharge, the radicals were oxidized by atmospheric oxygen. The bulk properties of the polymer were not influenced by the corona discharge.

Catoire et al. [18] studied the treatment of a low-density polyethylene (LDPE) film with corona discharge in the presence of air. The corona discharge created superficial ethylenic bonds and carbonyl groups, visible with ATR-IR spectroscopy. The amount of these groups varied inversely with the main power supply, similar to the results reported by us [19]. Formation of a thin waxy layer rich in carbonyl and ethylenic bond with almost total absence of ethylene groups characteristic of LDPE was also observed.

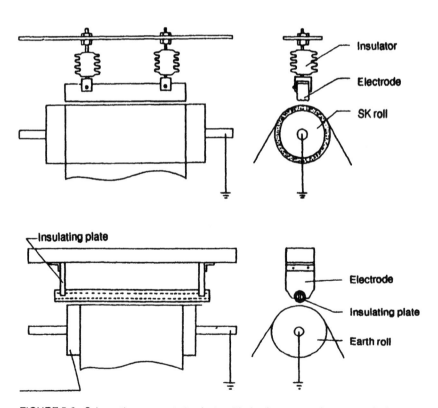

FIGURE 5.6. Schematic representation for two kinds of apparatus for corona discharge.

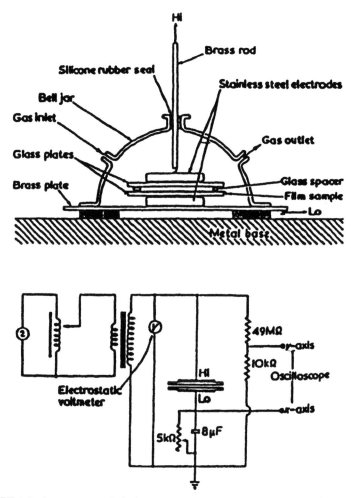

FIGURE 5.7. Arrangement of discharge treatment cell and electrical circuit diagram for discharge treatment system. (Source: Blythe et al. [21].)

Gerenser et al. [20] revealed through an XPS study on a corona-discharged, high-density polyethylene (HDPE) film that oxygen was incorporated into the surface in amounts ranging from 10 to 20 atom% depending on the power treatment levels. However, considerable chain scission occurred through the corona discharge at higher powers, giving water-soluble materials. These phenomena were in accordance with the fact that much of the oxidized substances could be easily removed from the corona-treated surface when washed with water.

Corona discharge treatment is widely used as an industrial process for enhancing the adhesive properties of polymer films. Blythe et al. [21] pointed out that although surface oxidation was observed from the XPS data even on PE corona-treated in inert gases, treatment in hydrogen resulted in no autoadhesion enhancement even though energy input into the film was more efficient than in air. Figure 5.7 shows the arrangement of the discharge treatment device they used. The same arrangement was used also by these authors to study the effects of film temperature during the treatment by corona discharge on a polypropylene surface [22].

5.4.4 UV Irradiation

Similarly to chemical vapor deposition (CVD), utilization of UV source provides a thin film deposit on the surface of various materials. White first reported in 1961 that 1,3-butadiene, in contact with metallic substrates, was polymerized upon irradiation with UV light [23]. It was indicated that a polymer deposit was formed selectively on illuminated areas of freshly prepared lead or tin films at a rate that was dependent on the nature of the underlying metal films. The deposition rate decreased at higher substrate temperatures. The UV light irradiation could provide a selective method for depositing very thin polymeric films on material surfaces in a patterned manner with high integrity. The typical apparatus for the process is illustrated schematically in Figure 5.8. The UV light source employed is commonly a medium-pressure or high-pressure mercury lamp that emits UV of wavelengths about 200 – 300 nm.

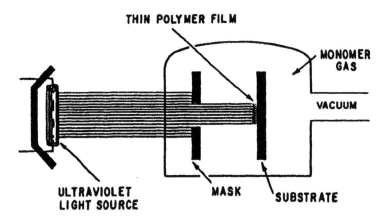

FIGURE 5.8. Schematic of UV surface-photopolymerization process of Kunz et al. [23].

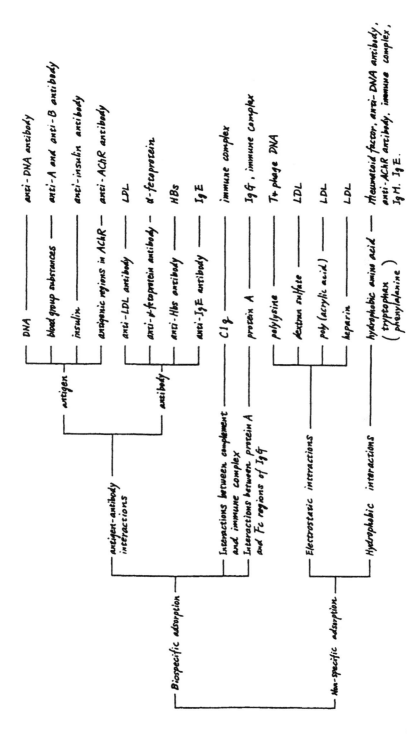

FIGURE 5.9. Classification of immunoadsorbents.

Biospecific adsorption	antigen-antibody interactions	antigen
		DNA — anti-DNA antibody
		blood group substances — anti-A and anti-B antibody
		insulin — anti-insulin antibody
		antigenic regions in AChR — anti-AChR antibody
		antibody
		anti-LDL antibody — LDL
		anti-α-fetoprotein antibody — α-fetoprotein
		anti-HBs antibody — HBs
		anti-IgE antibody — IgE
	Interactions between complement and immune complex	C1q — immune complex
	Interactions between protein A and Fc regions of IgG	protein A — IgG, immune complex
Non-specific adsorption	Electrostatic interactions	polylysine — T₄ phage DNA
		dextran sulfate — LDL
		poly (acrylic acid) — LDL
		heparin — LDL
	Hydrophobic interactions	hydrophobic amino acid (tryptophan phenylalanine) — rheumatoid factor, anti-DNA antibody, anti-AChR antibody, immune complex, IgM, IgE.

87

5.5 BIOLOGICAL MODIFICATIONS

Living cells have a physiologically active surface. Recently it became possible to create such surfaces on synthetic polymeric materials primarily by immobilization of active biomacromolecules. Indeed, there have been a large number of publications on protein immobilization. For this purpose, however, synthetic polymers must possess functional sites capable of binding proteins and other bioactive compounds at the surface region. Mostly, hydroxyl, carboxyl, and amino groups have been used for protein immobilization, but other functional groups such as aldehyde can also be utilized [24,25].

A recent trend of protein immobilization is to apply it for immunoadsorption. Since many immunologically mediated diseases are thought to be due to antibodies or immune complexes, attempts have been made to remove them from blood plasma using immunoadsorbents. Terman et al. [26] have studied an activated nylon microsphere with a surface immobilized with bovine serum albumin (BSA) which can remove anti-BSA antibodies selectively both *in vitro* and *in vivo*. The selective removal of circulating antibody specific for DNA was also studied with an immunoadsorbent consisting of DNA-cellulose conjugate incorporated into an agar gel [27]. Significant reduction in $_{ss}$DNA binding activity was observed over long periods after connection of the rabbit's circulations to the immunoadsorbent with only minimal changes in BSA binding during the same period. Little release of incorporated ^{125}I-labelled DNA from the column occurred during the procedure, as assayed in the blood and tissues of experimental animals. Regnault et al. studied a therapeutic immunoadsorption system with immobilized, anti-apolipoprotein B for lowering of plasma cholesterol concentration [28]. They covalently coupled several antibodies to Sepharose CL-4B activated with cyanogen bromide. The amount of antibodies released from immunoadsorbents could be minimized by treatment with a 0.005% glutaldehyde solution, but with an acceptable reduction rate of adsorption capacity. Under this condition, the immunoadsorption system could efficiently, specifically, and safely remove cholesterol from blood. Other possible combinations of immunoadsorbent and antibody to be removed are summarized in Figure 5.9.

5.6 REFERENCES

 1. Sumiya, K., T. Yasui, K. Nakamae and T. Matsumoto. 1982. *Nippon Setchaku Kyokaishi*, 18:345.
 2. Uyama, Y., H. Tadokoro and Y. Ikada. 1990. "Surface Lubrication of Polymer Films by Photoinduced Graft Polymerization," *J. Appl. Polym. Sci.*, 39:489–498.
 3. Iwamoto, R. 1986. *Jinkozoki* (written in Japanese), 15:1834.

4. Holmes-Farley, S. R., R. H. Reamey, T. J. McCarthy, J. Deutch and G. M. Whitesides. 1985. *Langmuir*, 1:266.

5. Zeronian, S. H., H.-Z. Wang and K. W. Alger. 1990. "Further Studies on the Moisture-Related Properties of Hydrolyzed Poly(ethylene terephthalate)," *J. Appl. Polym. Sci.*, 41:527−534.

6. Corretge, E., A. Kishida, H. Konishi and Y. Ikada. 1988. "Grafting of Poly(ethylene glycol) on Cellulose Surfaces and the Subsequent Decrease of the Complement Activation," *Polymers in Medicine III*, C. Migliatesi et al., eds., Amsterdam: Elsevier Science Publishers, pp. 61−72.

7. Kishida, A., K. Mishima, E. Corretge, H. Konishi and Y. Ikada. 1992. "Interactions of Poly(ethylene glycol)-Grafted Cellulose Membranes with Proteins and Platelets," *Biomaterials*, 13:113−118.

8. Suzuki, M. and Y. Ikada. 1981. *Radiat. Phys. Chem.*, 18:1207.

9. Klein, P. G., D. W. Woods and I. M. Ward. 1987. "The Effect of Electron Irradiation on the Structure and Mechanical Properties of Highly Drawn Polyethylene Fibers," *J. Polym. Sci. Part B Polym. Phys.*, 25:1359−1379.

10. Yasuda, H., M. O. Bumgarner and H. C. Marsh. 1975. "Ultrathin Coating by Plasma Polymerization Applied to Corneal Contact Lens," *J. Biomed. Mater. Res.*, 9:629−643.

11. Hollahan, J. R. and B. B. Stafford. 1969. "Attachment of Amino Groups to Polymer Surfaces by Radiofrequency Plasmas," *J. Appl. Polym. Sci.*, 13:807−816.

12. Bamford, C. H. and J. C. Ward. 1961. "Effect of High-Frequency Discharge on Surfaces of Solids," *Polymer*, 2:277−282.

13. Nuzzo, R. G. and G. S. Smolinsky. 1984. "Preparation and Characterization of Functionalized Polyethylene Surfaces," *Macromolecules*, 17:1013−1019.

14. Hall, J. R., C. A. L. Westerdahl, M. J. Bodnar and D. W. Levi. 1972. "Effect of Activated Gas Plasma Treatment Time on Adhesive Bondability of Polymers," *J. Appl. Polym. Sci.*, 16:1465−1477.

15. Bradley, A. and J. D. Fales. 1971. "Prospects for Industrial Applications of Electrical Discharge," *Chem. Tech* (April):232−237.

16. Tamaki, H. and S. Tatsuta. U.S. patent 4472467, Sep. 18, 1984.

17. Steinhauser, H. and G. Ellinghorst. 1944. "Corona Treatment of Isotactic Polypropylene in Nitrogen and Carbon Dioxide," *Angewandte Makromolekulare Chem.*, 120:177−191.

18. Catoire, B., P. Bouriot, O. Bemuth, A. Baszkin and M. Chevrier. 1984. "Physico-Chemical Modifications of Superficial Regions of Low-Density Polyethylene (LDPE) Film under Corona Discharge," *Polymer*, 6:766−772.

19. Iwata, H., A. Kishida, M. Suzuki, Y. Hata and Y. Ikada. 1988. "Oxidation of Polyethylene Surface by Corona Discharge and the Subsequent Graft Polymerization," *J. Polym. Sci.*, 26:3309−3322.

20. Gerenser, L. J., J. F. Elman, M. G. Mason and J. M. Pochan. 1985. "E.S.C.A. Studies of Corona-Discharge-Treated Polyethylene Surfaces by Use of Gas-Phase Derivatization," *Polymer*, 26:1162−1166.

21. Blythe, A. R., D. Briggs, C. R. Rance and V. J. I. Zichy. 1978. "Surface Modification of Polyethylene by Electrical Discharge Treatment and the Mechanism of Autoadhesion," *Polymer*, 19:1273−1278.

22. Briggs, D., C. R. Kendall, A. R. Blythe and A. B. Wootton. 1983. "Electrical Discharge Treatment of Polypropylene Film," *Polymer*, 24:47–52.

23. White, P. 1961. *Proc. Chem. Soc.*, 337; Wright, A. N. in *Polymer Surfaces*, pp. 155–184; Kunz, C. O., P. C. Long and A. N. Wright. 1972. *Polym. Eng. and Sci.*, 12:209.

24. Hayashi. T and Y. Ikada. 1990. "Protease Immobilization onto Polyacrolein Microspheres," *Biotech. and Bioengineering*, 35:518–524.

25. Hayashi, T. and Y. Ikada. 1990. "Lipoprotein Lipase Immobilization onto Polyacrolein Microspheres," *Biotech. and Bioengineering*, 36:593–600.

26. Terman, D. S., T. Tavel, D. Petty, A. Tavel, R. Harbeck, G. Buffalow and R. Carr. 1976. "Specific Removal of Bovine Serum Albumin (BSA) Antibodies *in vivo* by Extracorporeal Circulation over BSA Immobilized on Nylon Microcapsules," *J. Immunology*, 116:1337–1341.

27. Terman, S., I. Stewart, J. Robinette, R. Carr and R. Harbeck. 1976. "Specific Removal of DNA Antibodies *in vivo* with an Extracorporea Immuno-Adsorbent," *Clin. Exp. Immunol.*, 24:231–237.

28. Regnault, V., C. Rivat, P. Marcillier, M. Pfister, J. P. Michaely, J. Didelon, F. Schooneman, J. F. Stoltz and M. Siadat. 1990. "Study of Parameters Involved in Specific Immunoadsorption of Apolipoprotein B," *Int. J. Artif. Organs*, 13:760–767.

CHAPTER 6

Surface Analysis of Modified Polymers

6.1 INSTABILITY OF POLYMER SURFACES

It is difficult to rigorously define a polymer surface, since a "surface" includes polymer chains both in the parallel region of the plane and at a certain depth from the outermost surface, as was mentioned in the preceding chapter. The term "surface" should, in principle, denote the monolayer plane that separates the vapor or the vacuum phase from the condensed liquid or the solid phase. In other words, the surface layer should be only monomolecular in thickness. The definition of the surface as well as the depth of the surface region might therefore depend on an individual researcher's viewpoint. Furthermore, polymer chains at the surface are driven toward a thermodynamically lower energy state and are constantly changing their positions, as eels squirming in a crowd. The fact that polymer chains do not maintain a fixed and constant configuration even at ambient temperature makes the surface analysis much more complicated and ambiguous.

It is known that polymer surfaces oxidized by corona or glow discharge treatment become hydrophobic again when stored in air or vacuum at room temperature. This is most probably due to the gradual overturn or the reorientation of polar groups into the hydrophobic bulk phase to reduce the overall free energy [1]. Piao et al. [2] studied the aging of plasma-treated PET and PE surfaces under different conditions by measuring the change in the contact angle of these surfaces with time. The rate of this change depended on both the storage temperature and the surrounding environment. For instance, the contact angle of plasma-treated PET film increased from 5 to 53°C during storage in air at 25°C for seven days, while it increased to 59°C within one hour when heated in air to 105°C. The plasma-treated PET surfaces maintained their hydrophilicity if stored in water or aqueous alkaline solution, or in air at extremely low temperatures such as −25°C. Yasuda et al. [3] extensively investigated the dynamics of the surface configuration in response to changes in the environmental condition using a CF_4 plasma surface labeling technique. They observed a noticeable change of the surface configuration after immersion in water. This change could be expressed by the following general equation, which describes the phenomenon of diffusing species chemically reacting with a medium

$$A_t = A_0 \, t^{-k} \tag{6.1}$$

where A is a parameter that describes a surface property, A_0 and A_t are the corresponding values of A for a sample that was immersed in water for times zero and t, respectively, and the value k can be calculated from the initial linear portion of a log-log plot. The parameter k may be taken as describing the surface dynamic mobility of polymers and is related to the ease of rotational and diffusional migrations of hydrophobic moieties from the surface into the bulk of a film. They selected a variety of polymers having a wide range of glass transition temperature (T_g) from $-125\,°C$ (PE) to $105\,°C$ (PMMA), and concluded that, as a consequence of the molecular motion, the apparent transition temperature T_s, at which an abrupt change in the rate of surface configuration alteration occurs, was below the T_g of these polymers [4].

Even long grafted chains might be buried inside the polymer bulk phase, depending on the surrounding environment. Under special conditions polymer molecules would migrate together with the graft layer at a temperature higher than the T_g of the polymeric materials. Instability of surface graft layers was discussed by Ratner et al. [5] for poly(dimethyl siloxane), polyesterurethane, and polyethylene grafted with 2-hydroxyethyl methacrylate (HEMA), acrylamide, and ethyl methacrylate polymers. The hydrophilic surfaces derived from the modification, either by introducing polar groups or by surface graft polymerization, were lost upon heating at elevated temperatures far above T_g. Once the graft polymers were buried inside the polymer bulk phase, it was not easy to restore the hydrophilic nature of the surface.

6.2 METHODS FOR ANALYZING FUNCTIONAL GROUPS ON POLYMER SURFACES

In spite of the recent and dramatic progress in surface analysis, there still remain unsolved problems relating to the surface characteristics of polymers. For instance, Fourier-transform infrared spectroscopy coupled with attenuated total reflection (ATR FT-IR), Auger spectroscopy, and X-ray photoelectron spectroscopy (XPS) are able to provide valuable information regarding the constituent elements and chemical structures near the surface region. However, in general, these analyses do not provide information on functional groups dangling at the outermost surface in a quantitative manner. This may be because the absolute amount of the functional groups present at the outermost surface is too small to be quantitatively determined even by such sophisticated analytical methods. Nevertheless, it has been demonstrated that very small amounts of the functional groups newly created at or close to the surface of LDPE films oxidized with concentrated chromic acid followed by further oxidation with aqueous nitric acid [6]. PVA films reacted with hexamethylene diisocyanate can *also* be *successfully* determined by fluorescence spectroscopy [7]. Leclercq et al. [8] have determined the surface density of the functional groups generated on corona-treated poly(ethylene terephthalate) by adsorption of radio-labeled calcium ions. The surface concentration of the hydroxyl groups of cellulose and poly(vinyl alcohol) films was successfully determined utilizing the chemical derivatization technique [9].

The chemical derivatization technique has also been employed for surface analysis by XPS. Although XPS spectra provide sufficient information for the identification of surface elements and functional groups, the analysis often fails to quantify them because of indistinguishable chemical shifts between different functional groups. For instance, the C_{1S} binding energies for $C-O-H$, $C-O-C$, and $C-O-O-H$ groups all fall in the range of 286 to 287 eV. In addition, these functional groups exhibit core-level spectra with quite a large FWHM (full width at half maximum), typically in the order of $1.2-1.7$ eV. Nuzzo and Smolinsky [10] oxidized the surface of polyethylene film with a combination of plasma and chromic acid treatments, and analyzed functional groups by XPS and other analytical methods after selectively derivatizing the surface groups by trifluoroacetic anhydride and radiolabeled reagents. Typical chemical derivatization reactions are summarized in Table 6.1. [11−16]. The derivatization reactions utilized for XPS studies should meet the following requirements:

(1) The reagents for chemical derivatization should react quantitatively and selectively with the functional groups existing on the substrate surface in a short time.

(2) The excess of the reagents should be easily removable from the substrate surface.

(3) The derivatized groups should not decompose during the XPS measurement.

(4) The derivatized groups should have significant chemical shifts and intensities.

TABLE 6.1. XPS—Chemical Modification of
Functional Groups on the Polymer Surface.

Functional Groups	Reagents Used for Modification	Derivatives	References
$-\overset{\shortmid}{C}=\overset{\shortmid}{C}-$	Br$_2$/CCl$_4$	$\overset{\overset{\text{Br}}{\shortmid}}{-\overset{}{C}}-\overset{\overset{\text{Br}}{\shortmid}}{\underset{}{C}}-$	[11]
	Hg(OAc)$_2$ F$_3$CCH$_2$-OH	CF$_3$CH$_2$-O Hg(OAc) $-\overset{\shortmid}{C}-\overset{\shortmid}{C}-$	[12]
$\overset{\shortmid}{C}$-OH	(CF$_3$CO)$_2$O	$-C-O-CO-CF_3$	[12]
	Ti(acac)$_2$OPr$^i_2$	$-CH_2OTi(acac)OPr^i$	
	CBr$_3$CO$_2$H C$_6$H$_{11}$NCNC$_6$H$_{11}$	$-O_2CCBr_3$	[13]
			[12]
$C=\overset{\shortmid}{C}$-OH	ClCH$_2$COCl	$C=\overset{\shortmid}{C}$-OCOCH$_2$Cl	[13]
$-CH_2-C\overset{\nearrow O}{\diagdown}$	Br$_2$	$-CBr_2-C\overset{\nearrow O}{\diagdown_O}$	[13]
$-\overset{\shortmid}{C}=O$	C$_6$F$_5$NHNH$_2$	$-\overset{\shortmid}{C}=NNHC_6F_5$	[14]
$-C-OH$ $\overset{\shortmid}{O}$	NaOH	$-CO_2^-Na^+$	[14]
	BaCl$_2$	$(-CO_2^-)_2Ba^{2+}$	[15,16]
	CF$_3$CH$_2$OH C$_6$H$_{11}$NCNC$_6$H$_{11}$	$-COOCH_2CF_3$	[12]
	(1) KOH (2) C$_6$F$_5$CH$_2$Br	$-COOCH_2C_6F_5$	[12]
	SO$_2$	$-C-OSO_2OH$	[13]
$-NH_2$	CH$_3$CH$_2$SCOCF$_3$	$-NHCF_3$	[16]
	C$_6$F$_5$CHO	$-N=CHC_6F_5$	[12]
	C$_2$H$_5$S-COCF$_3$	$-NH-COCF_3$	[16]

94

6.3 CONTACT ANGLE METHODS

The contact angle measurement is a very old method used for surface analysis. However, it is still more widely used than the other techniques because of its ease in measurement and low cost in instrumentation. In addition, it should be stressed that the contact angle gives valuable information on the properties of the outermost thin layer of polymers, such as hydrophilicity-hydrophobicity balance, overturn of functional groups, and microscopic roughness at the surface. It should further be pointed out that contact angle can characterize polymer surfaces within 1 nm depth, while XPS provides information on the elemental ratios and functional groups of polymer surfaces within 5 nm [17]. Thus, one can say that polymer surfaces can be best characterized utilizing a combination of both classical and modern instruments.

There are numerous methods to measure contact angle: the sessile drop and the adhering gas bubble methods, the Wilhelmy gravitational method, the capillary rise method at a vertical plate, the tilting plate method, the reflection method, etc. As a detailed description is available in the literature [18,19], only the relatively newer methods will be described below.

6.3.1 Laser Beam Contact Angle Measurement

The laser beam contact angle measurement was proposed in 1982 independently by Israel [20] and Kishi [21]. The instrument consists of optical components including a laser source, a filter, a sample stage, and a screen on an optical bench, as shown in Figure 6.1. The laser beam reflected from the solid-liquid-air interface is projected on the screen normal to the incident beam at an angle α, which is related to the contact angle θ by Equation (6.2):

$$\theta = 90° - \alpha \qquad (6.2)$$

As the projected beam on the screen is clear and large, one can easily determine the contact angle within $\pm 0.2°$ with the naked eye using a protractor drawn on the screen. The beam commonly employed is a HeNe laser of 632.8 nm wavelength, with less than 1 mW power. A typical photograph demonstrating the reflected laser beam on the screen is shown in Figure 6.2 for a cellulose film. In this example α is 65° and θ is 25°, and the laser beam projected on the protractor scale is large and clear enough to read the contact angle within $\pm 0.1°$. Thus, with respect to accuracy and the cost of the equipment, the laser beam method appears to be more favorable than the conventional telescopic sessile drop technique. It should be further noted that a water droplet as small as 0.05 μl can be used for the laser beam method, so that it is possible to determine the contact angle on

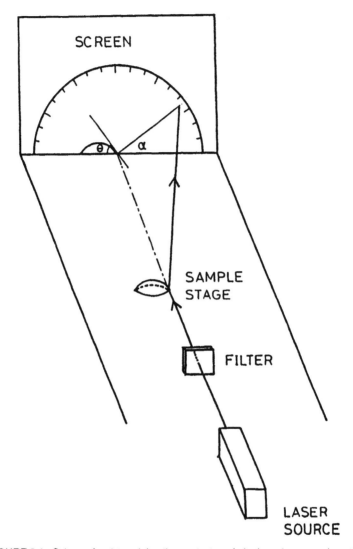

FIGURE 6.1. Scheme for determining the contact angle by laser beam goniometry.

a very small surface of specimen such as monofilament fiber. The drawback of this laser beam method is the difficulty in determining contact angles around 90° or larger, and in applying it to the inverted bubble method. Israel et al. [22] studied the effect of droplet size on the stationary contact angles of PMMA and glass using their apparatus, and found that the contact angle decreased with decreasing droplet size. They noted, however, that the change in the contact angle was as small as 2.6° even when the droplet size was enlarged from 0.2 to 10 μl.

A comparison was made between the contact angles measured by the telescopic and the laser beam goniometry methods for polymer films of different wettabilities [23]. As Figure 6.3 shows, an excellent correlation was found between the two methods, with a small experimental error associated with the laser beam contact angle compared to that of the telescopic contact angle method. The contact angles measured with the Wilhelmy plate method are also given in Figure 6.3.

6.3.2 Wilhelmy Plate Method

Many polymer systems display some degree of contact angle hysteresis, which is defined as the difference between the advancing and the receding contact angle. Extensive studies on contact angle hysteresis were reported by Smith et al. employing the Wilhelmy technique [24,25]. The equipment used in their studies consisted of a mechanical testing device, a balance system, and an x-y plotter, as shown in Figure 6.4. The result on polybutadiene is shown in Figure 6.5. As the crosshead is raised, the sample is immersed in the test fluid. When the sample first touches the fluid surface, a meniscus is formed and, as immersion of the sample into the liquid continues, the contact angle levels off, giving a constant slope on the x-y recorder, which is a measure of the advancing contact angle. After immersion to $2.5 - 3.0$ cm, the process is reversed, the wetting liquid is lowered, and a constant slope due to a receding angle is observed on the x-y recorder.

FIGURE 6.2. Laser beam projected on a screen for a water droplet on a cellulose film.

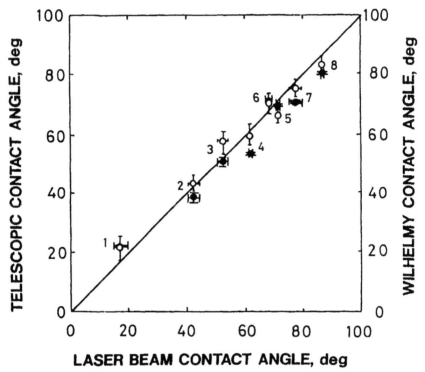

FIGURE 6.3. Comparison of the contact angles obtained from the telescopic (○), and the Wilhelmy plate technique (●) with the laser beam goniometry. (Source: Uyama et al. [23].)

They performed the measurements keeping the mechanical tester and balance insulated and maintained a constant temperature (20°C) and humidity (30% RH).

Generally, a straight-line approximation of the advancing and receding slopes is made, and extrapolated to zero depth of immersion to eliminate the buoyancy factor B_f from Equation (6.3):

$$B_f = V\varrho g/p\gamma \qquad (6.3)$$

where m is the mass of the slide as measured via electrobalance; g is the local gravitational force; p is the perimeter of sample; γ is the surface tension of wetting liquid (for water γ = 72.6 dyne/cm at 20°C); V is the volume of immersed sample at a particular depth; and ϱ is the density of wetting liquid. The basic equations for determining the advancing (θ_a) and the receding contact angles (θ_r) are given by:

$$F_a = \gamma L \cos \theta_a \qquad (6.4)$$

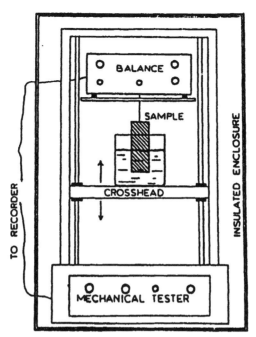

FIGURE 6.4. Schematic of Wilhelmy plate equipment. A mechanical tester is used to raise and lower wetting liquid over sample interface at 40 mm/min. Wetting forces are measured on an electrobalance [24].

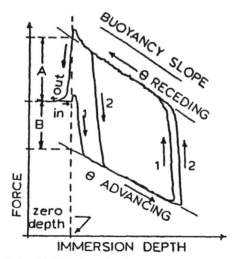

FIGURE 6.5. A typical stable Wilhelmy plate hysteresis loop for a polybutadiene sample. The buoyancy slope is parallel in both the advancing and the receding cases. The sample was taken through two successive cycles to demonstrate reproducibility. Displacement A is used to compute the receding angle; displacement B is used to compute the advancing angle. The data were obtained at the point of zero depth of immersion so the buoyancy effect can be neglected [25].

$$F_r = \gamma\, L \cos \theta_r \qquad\qquad (6.5)$$

where F_a and F_r represent the observed forces at zero depth immersion of samples at the advancing and the receding cycle, respectively, and L is the peripheral of the immersed sample.

The hysteresis of contact angle [26] can be classified into two groups:

(1) *Thermodynamic hysteresis*, where the hysteresis curve is reproducible over many cycles and is independent of time and frequency.

(2) *Kinetic hysteresis*, where the hysteresis curve changes with time and circumstances.

The hysteresis observed for the Wilhelmy plate measurement on the poly(butadiene) surface is mainly due to thermodynamic hysteresis.

Another example of the Wilhelmy plate measurement is shown in Figure 6.6 for a PHEMA hydrogel, for which multiple cycles of measurement were not possible because of kinetic hysteresis. The Wilhelmy plate hysteresis loop depended on time and position, since the dried polymer specimen swelled in water, changing both in weight and wettability with time.

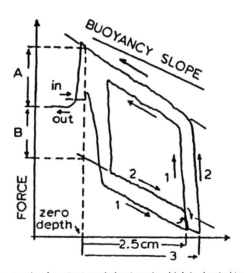

FIGURE 6.6. An example of contact angle hysteresis which is due to kinetic or time-dependent effects. In this particular case an unhydrated poly(hydroxyethyl methacrylate) gel was immersed in water and the Wilhelmy plate experiment carried out. The gel swells during the course of the experiment. As it swells, its mass, dimensions, and wettability change. The result is that the recorded force increases with time in contact with water. Note that on the second cycle the gel was immersed a short distance beyond the immersion distance for the first cycle, i.e., dry gel is seeing water for the first time. The plot on cycle 2 is parallel to that of cycle 1 for that portion of the gel which was still initially dry [25].

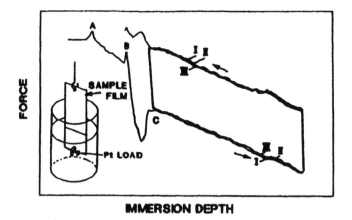

IMMERSION DEPTH

FIGURE 6.7. Hysteresis loop obtained for a HDPE film by the Wilhelmy plate technique at an immersion rate of 10 mm/min. I – III corresponds to the first, second, and third cycles of measurement [23].

The Wilhelmy plate contact angle method can give further information on the surface features of polymeric materials, for instance, surface roughness, heterogeneity, deformation, swelling and penetration of liquid, and surface reorientation and mobility. However, the Wilhelmy plate method has one disadvantage in that both surfaces of the sample plate must be identical. Otherwise, the x-y plot would be difficult to interpret. Andrade et al. coated the polymer specimen on both sides of a glass plate [25].

Currently, commercial apparatus are available for the measurement of Wilhelmy contact angle. For instance, in Japan, an apparatus (Automated System for Contact Angle Measurement, ST-1S type) equipped with a microcomputer for analyzing advancing and receding contact angles and related surface energetics has been manufactured by Simadzu Inc., Kyoto, Japan. The design of this apparatus, as well as the principle of determining the contact angle, is similar to that described by Andrade et al. [25]. Since most of the polymeric materials have a density less than 1.0, it is often difficult to immerse the polymer films deeply into water. Instead of coating the polymer on a heavy glass plate, a polymer film can be immersed directly in water by attaching a load to the bottom of the film, as illustrated in Figure 6.7. An example of the results obtained for a HDPE film against water is also shown in Figure 6.7, where the sample film with a Pt load was immersed at a contact speed into purified water [23]. When the Pt load and subsequently the HDPE film touched the water surface, the meniscus was formed in both cases (A and B in Figure 6.7). Until the Pt load was completely immersed in the water, the curve of the chart was meaningless (from B to C in Figure 6.7). Thereafter, as the sample film was further immersed in the water, the surface tension of the water on the film became constant, but the film buoyancy linearly decreased, resulting in a constant slope on the *x-y* recorder.

6.4 SPECTROSCOPIC METHODS

Although a variety of analytical methods are currently available for characterizing polymer surfaces, physical analytical tools are the most widely employed. These techniques include XPS, SIMS, Fourier-transform infrared spectroscopy (FT-IR) coupled with attenuated total reflection (ATR), scanning electron microscope (SEM), and energy dispersive X-ray. Several of these modern physical techniques such as XPS and SIMS use bombarding particles such as photons, electrons, neutrons, or ions on the surface of the sample. The energy of emitted particles gives a variety of surface information, including identification of atoms and chemical bonds present in the very thin surface layer. The possible combinations of probes for incident and emitted particles are as follows (see Figure 6.8):

(1) The incident energy probe is an electron and the emitted energy is also an electron (auger electron); [auger electron spectroscopy (AES)].

(2) The incident and emitted energies are both ions [Rutherford back-scattering (RBS)].

(3) The incident and emitted energies are both X-rays [X-ray fluorescence spectroscopy (XRF)].

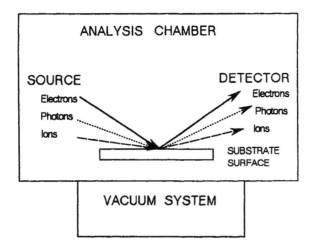

FIGURE 6.8. The combinations of probes for incident and emitted particles for physical spectroscopic analysis.

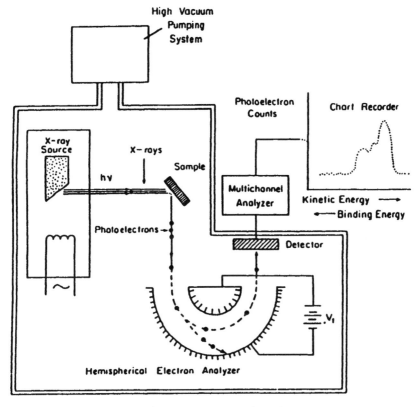

FIGURE 6.9. Schematic diagram of an XPS system [27].

(4) The incident probe is an X-ray, whereas the emitted energy is an electron [X-ray photoelectron spectroscopy (XPS)].

(5) The incident energy is an electron, whereas the emitted energy is an X-ray [electron microprobe analysis (EMA)].

(6) The incident energy is an ion beam, and the target ion is emitted [secondary ion mass spectroscopy (SIMS)].

6.4.1 XPS Analysis

XPS is often called ESCA, which is the abbreviation for electron spectroscopy for chemical analysis. When the surface of substrate materials including the deep region of a bulk phase is subjected to electromagnetic radiation, low-energy electrons are emitted by a process called photoionization. The principle of XPS consists of counting the number of photoelectrons emitted at specific energies by X-ray bombardment (Figure 6.9)

[26]. Although X-rays can penetrate deeply into the polymer bulk phase, the mean free path of photoemitted electrons through the polymer surface region depends on their kinetic energy. The photon sources employed in XPS study are MgKα, AlKα, TiKα, and CrKα. Among them, MgKα has the lowest energetic strength and CrKα the highest. If it is necessary to increase the detection level of XPS spectra, a higher energy source can be used but the degree of resolution becomes poorer. The most commonly used source for analyzing polymer surfaces is MgKα, which provides information on the sampling depth of about 5 nm with good resolution for most polymeric materials.

Among the various probe-based analytical methods described above, XPS is most often employed for surface analysis, because it has the following advantages:

(1) It can be applied to almost all polymeric materials without severely damaging the surface structure during the measurement.

(2) It can identify almost every element present in the surface region whose chemical composition may be virtually different from that of the bulk.

(3) It can determine quantitatively any elements present at the surface layer, except hydrogen and helium.

(4) It can reveal to some extent the surface electrical properties from surface charging effects.

(5) It can evaluate the nature of molecular bonding, the chemical environment, and the oxidation states for most elements at the surface.

In addition, XPS can determine the compositional gradient extending from the surface into the bulk phase of the polymer by changing the incident angle of the X-ray source, as shown in Figure 6.10. In practice, the positioning of the X-ray source and the analyzer remains unchanged, but only the angle of the sample holder, which has a specific angle cutting, is changed. The mean free path of the emitted photoelectron is simply varied by the factor of cos θ, where θ is the angle between the inclined sample and the incident X-ray. The depth profiling analyses are performed using special algorithms. A number of articles describe the methods to convert the angular-dependent XPS data into depth profiles [27–30].

The XPS study provides information not only on the chemical components and their depth profiles but also on the nature of bondings derived from the chemical shifts. Indeed, many investigators have utilized the chemical shifts to identify newly produced functional groups on polymer surfaces. For instance, the oxidation of the poly(1-trimethylsilyl-1-propyne) membrane, which has the highest gas permeability among conventional polymers, was studied using the XPS technique [31]. This

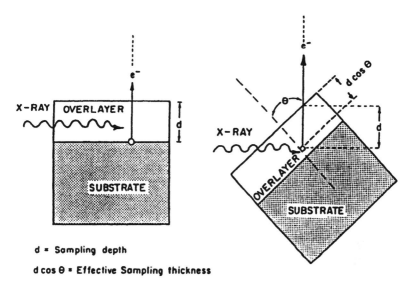

d = Sampling depth

d cos θ = Effective Sampling thickness

FIGURE 6.10. Decrease in the effective sampling depth as the angle of the specimen (θ) is increased with respect to the normal position (X-ray source and detector in fixed positions) [27].

membrane became irreversibly hydrophilic only by being heated in air or immersed in water at room temperature. The XPS measurement revealed that the hydrophilic surface was produced by spontaneous oxidation, as is indicated in Figure 6.11 by the increasing intensities of the high binding energy components in the C_{1s} core-level spectra with the treatment time.

6.4.2 SIMS

The principle of static secondary ion mass spectroscopy (SIMS) resembles somewhat that of XPS, but the source of bombardment and the particles detected are quite different from those of XPS. In SIMS, an accelerated ion beam (the primary beam) is impinged upon a test surface, and the atomic and molecular fragments emitted upon bombardment of the high-energy ion beam are detected as a mass spectrum similar to the method employed in conventional mass spectroscopy (MASS). The energy of the incident ion beam is transferred to atoms and molecules in the surface region to disrupt interatomic attractions and break covalent bonds. Thus, in contrast to XPS, the surface of the test specimen usually suffers significant etching, leading to erosion of the surface layer by sputtering. The relatively large abundance of emitted species provides a direct measure of the chemical composition and structure of the surface layer that has been sputtered by the ion beam. This process is illustrated schematically in Figure 6.12. Emitted species are in the form of neutral fragments in variously excited states, or in the form of singly or doubly charged positive and negative ion fragments. The molar ratio of ionized to neutral species from the same specimen can vary, depending on the energies and sources of primary ions. A widely utilized technique is to collect and analyze the positively and

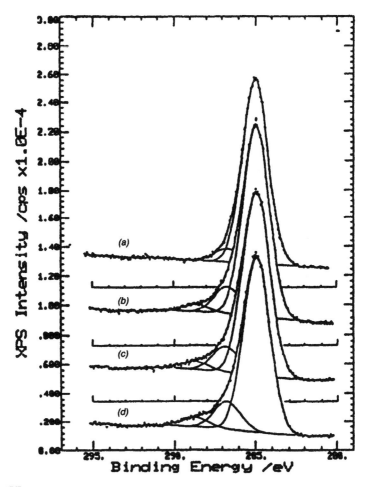

FIGURE 6.11. High-resolution XPS spectra of the C_{1s} peak for PTMSP: (a) virgin; (b) heat-treated in air at 110°C for 12 hours; (c) heat-treated in air at 110°C for 24 hours; and (d) heat-treated in air at 110°C for 48 hours [31].

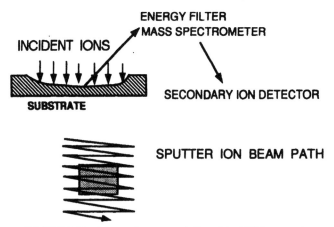

FIGURE 6.12. Schematic representation of the SIMS apparatus.

negatively charged fragments (secondary ions) in a mass spectrometer (typically of the quadrapole type) to provide, in sequential experiments, the positive and negative SIMS spectra, respectively. A detailed description of the process can be found in an article, for instance, written by Castner and Ratner [32].

6.5 ATR FT-IR

The method of attenuated total reflection (ATR) for surface analysis was first proposed independently by Harrick [33] and Fahrenfort [34]. However, the infrared (IR) spectroscopy of the early 1960s did not give sufficient information on the nature of the surface, and hence the ATR method was applied only for qualitative analysis. Recently, improvements in IR spectroscopy have led to high accuracy by utilizing Fourier transformation with the aid of computers in the analysis of the spectra. The principle of the measurement is quite simple. The infrared light reflects through the internal reflection element (IRE), but is attenuated by absorption of chemical moieties existing in the surface region of the sample polymer. The infrared absorption is analyzed by a photometer, similar to a conventional IR analysis. Figure 6.13 shows a schematic diagram of total reflection of the light beam in ATR-IR measurements. The materials commonly used for IRE are KRS-5, Ge, Al_2O_3, and Si. As the number of reflections depends on the angle of the incident beam, it is necessary to adjust it before performing the analysis if precise information is needed from the ATR FT-IR measurement.

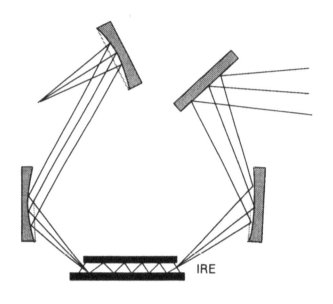

Wilks type

sample
mirrors

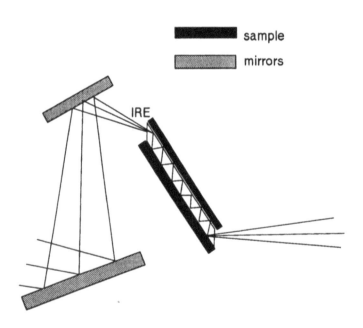

Harrick type

FIGURE 6.13. A brief diagram of ATR-IR measurement.

6.6 REFERENCES

1. Ikada, Y., T. Tatsunaga and M. Suzuki. 1985. *Nippon Kagaku Gakkaishi*, (6):1079.

2. Piao, D.-X., Y. Uyama and Y. Ikada. 1991. "Aging of Plasma Treated Polymers," *Kobunshi Ronbunshu*, 48:529–534.

3. Yasuda, T., T. Okuno, K. Yoshida and H. Yasuda. 1988. "A Study of Surface Dynamics of Polymers. II. Investigation by Plasma Surface Implantation of Fluorine-Containing Moieties," *J. Polym. Sci: Part B, Polym. Phys.*, 26:1781–1794.

4. Yasuda, H., E. F. Charlson, E. M. Charlson, T. Yasuda, M. Miyama and T. Okuno. 1991. "Dynamics of Surface Property Change in Response to Changes in Environmental Conditions," *Langmuir*, 7:2394–2400.

5. Ratner, B. D., P. K. Weathersby, A. S. Hoffman, M. A. Kelley and L. H. Shaypen. 1978. "Radiation-Grafted Hydrogels for Biomaterials: Applications as Studied by the ESCA Technique," *J. Appl. Polym. Sci.*, 22:643–664.

6. Rasmussen, J. R., E. R. Stedronsky and G. M. Whitesides. 1977. "Introduction, Modification, and Characterization of Functional Groups on the Surface of Low-Density Polyethylene Film," *J. Amer. Chem. Soc.*, 99:4736–4745.

7. Caro, J. R., C. S. Paik-Sung and E. W. Merril. 1976. "Reaction of Hexamethylene Diisocyanate with Poly(vinyl alcohol) Films for Biomedical Applications," *J. Appl. Polym. Sci.*, 20:3241–3246.

8. Leclercq, B., M. Sotten, A. Baszkin and L. Ter-Minassian-Saraga. 1977. "Surface Modification of Corona Treated Poly(ethylene terephthalate) Film: Adsorption and Wettability Studies," *Polymer*, 18:675–680.

9. Matsunaga, T. and Y. Ikada. 1980. In *Modification of Polymers*, E. Charles et al., eds., ACS Symp. Ser. 121, pp. 391–406.

10. Nuzzo, R. G., and G. Smolinsky. 1984. "Preparation and Characterization of Functionalized Polyethylene Surfaces," *Macromolecules*, 17:1013–1019.

11. Briggs, D., D. M. Brewis and M. B. Konieczko. 1977. *J. Mater. Sci.*, 12:429.

12. Everhart, D. S. and C. N. Reilley. 1981. "Chemical Derivatization in Electron Spectroscopy for Chemical Analysis of Surface Functional Groups Introduced on Low-Density Polyethylene Film," *Anal. Chem.*, 53:665–675.

13. Briggs, D. and C. R. Kendall. 1982. *Int. J. Adhesion, Adhesives*, 2:13.

14. Briggs, D. and C. R. Kendall. 1979. "Chemical Basis of Adhesion to Electrical Discharge Treated Polyethylene," *Polymer*, 20:1053–1054.

15. Czuha, J. M. and W. M. Riggs. 1975. "X-Ray Photoelectron Spectroscopy for Trace Metals Determination by Ion-Exchange Absorption from Solution," *Anal. Chem.*, 47:1836–1838.

16. Bradley, A. and M. Czuha, Jr. 1975. "Analytical Methods for Surface Grafts," *Anal. Chem.*, 47:1838–1840.

17. Holmes-Farley, S. R. and G. M. Whitesides. 1987. "Reactivity of Carboxylic Acid and Ester Groups in Functionalized Interfacial Region of 'Polyethylene Carboxylic Acid' (PE-COOH) and Its Derivative: Differentiation of the Functional Groups into Shallow and Deep Subsets Based on a Comparison of Contact Angle and ATR-IR Measurements," *Langmuir*, 3:62–76.

18. Neumann, A. W. and R. J. Good. 1976. In *Surface and Colloid Science, Vol. 11*, R. J. Good and R. R. Stromberg, eds., Plenum, pp. 31–91.

19. Adamson, A. W. 1976. *Physical Chemistry of Surfaces*. John Wiley, pp. 1–45; Vold, R. D. and M. J. Vold. 1983. *Colloid and Interface Chemistry*. Addison-Wesley, pp. 115–149.

20. Israel, S. C. U.S. patent 718865, 1984.

21. Kishi, N. Japan patent 6160370, Dec. 20, 1986.

22. Israel, S. C., W. C. Yang, C. H. Chae and C. Wong. 1989. "Characterization of Polymer Surfaces by Laser Contact Angle Goniometry," *ACS Polym. Prep.*, 30(1):328–329.

23. Uyama, Y., H. Inoue, K. Ito, A. Kishida and Y. Ikada. 1991. "Comparison of Different Methods for Contact Angle Measurement," *J. Colloid Interface Sci.*, 141(1):275–279.

24. Smith, L., C. Doyle, D. E. Gregonis and J. D. Andrade. 1982. "Surface Oxidation of cis-trans Polybutadiene," *J. Appl. Polym. Sci.*, 26:1269–1276.

25. Andrade, J. D., L. M. Smith and D. E. Gregonis. 1985. "The Contact Angle and Interface Energetics," in *Biomedical Polymers, Vol. 1*, J.D. Andrade, ed., New York: Plenum, pp. 249–292.

26. Johnson, R. E. and R. Dettre. 1969. In *Surface Colloid Sci., Vol. 2*, E. Matijevic, ed., Wiley, p. 85.

27. Ratner, B. D. and B. J. McElroy. 1986. "Electron Spectroscopy for Chemical Analysis: Applications in the Biomedical Sciences," *Spectroscopy in Biomedical Sciences*, R. M. Gendreau, ed., CRC Press, pp. 107–140; Ratner, B. D., T. A. Horbett, D. Shuttleworth and H. R. Thomas. 1981. *J. Colloid Interface Sci.*, 83:630.

28. Peeling, J., M. Jazzar and D. T. Clark. 1982. "An ESCA Study of the Surface Ozonation of Polystyrene Film," *J. Polym. Sci., Polym. Chem. Ed.*, 20:1797–1805.

29. Clark, D. T. and H. R. Thomas. 1978. "Application of ESCA to Polymer Chemistry XVII. Systematic Investigation of the Core Levels of Simple Homopolymers," *J. Polym. Sci., Polym. Chem. Ec.*, 16:791–820.

30. Kodama, M., K. Kuramoto and I. Karino. 1987. "ESCA and FTIR Studies on Boundary-Phase Structure between Blend Polymers and Polyamide Substrate," *J. Appl. Polym. Sci.*, 34:1889–1900.

31. Hata, Y., A. Kishida, Y. Ikada, T. Masuda and T. Higashimura. 1986. *Polym. Prep. Japan*, 35:3138.

32. Castner, D. G. and B. D. Ratner. In *Surface Characterization of Biomaterials*, B. D. Ratner, ed., Elsevier.

33. Harrick, N. J. 1960. "Surface Chemistry from Spectral Analysis of Totally Internally Reflected Radiation," *J. Phys. Chem.*, 64:1110–1114.

34. Fahrenfort, J. 1961. *Spectrochim. Acta.*, 17:698.

CHAPTER 7

Surface Grafting

As was demonstrated in Chapter 2, an extremely hydrophilic or hydrophobic surface will exhibit high lubricity. From a practical point of view, the extremely hydrophilic polymer surface is more attainable, because the most hydrophobic surface currently available has a water contact angle around 120°, which is much lower than 180°. Although oxidation of polymeric materials either by plasma treatment or by chemical agents readily produces hydrophilic surfaces (as was shown in Chapter 5) the obtained lubricity is not sufficient and often lasts only for a limited period of time. As is readily anticipated from the previous chapters, it is likely that a polymeric material having a surface that is capable of retaining a large amount of water should exhibit a high degree of lubricity when wetted. Such a material can be produced by incorporating water-soluble chains onto the surface.

In principle, there are two possible methods for incorporating water-soluble polymers: either blending the water-soluble polymer into the surface region of the substrate, or grafting the water-soluble polymer onto the surface of the substrate. Another method — such as physical coating of a water-soluble polymer layer on the substrate surface — is not a good method because of the gradual desorption of the polymer. Here, "physical coating" means the simple application of hydrophilic polymers onto the substrate surface. It is also likely that the water-soluble polymers incorporated onto the surface by physical mixing will undergo elution with water. It follows that the most appropriate way to render a material surface lubricious in the presence of water is to covalently immobilize water-soluble chains onto the substrate surface — that is, surface grafting.

Although surface grafting seems an interesting surface modification method with a number of possible applications, little attention has been directed toward this method. The fact that there are only a limited number of studies in this field is probably due to the difficulty associated with the reaction and the assessment of the products. Surface grafting can be achieved by two different means: a coupling reaction of an existing water-soluble polymer to the surface, or graft polymerization of a water-soluble monomer onto the substrate surface. The latter method has been explored much more extensively.

7.1 GRAFTING BY COUPLING REACTION

The grafting reaction includes all types of chemical reactions using a reactive group on both the substrate and the polymer to be grafted.

Hommel et al. [1] studied the influence of the grafting ratio on the configurations of PEO chains grafted on silica. The grafting reaction was performed by dispersing silica in excess of PEO, followed by careful degassing and heating at 230°C for 16 hours under a nitrogen atmosphere. Heating allowed direct esterification of the silanol groups with the hydroxyl end groups of PEO.

$$SiOH + ROH \rightarrow SiOR + H_2O \qquad (7.1)$$

We have grafted dextran onto a surface of ethylene-vinyl alcohol copolymer (EVAL) through a chemical coupling reaction [2]. Figure 7.1 shows the scheme of the coupling reaction of dextran with or without primary amino groups, onto the EVAL surface. The coupling reaction took place via a urethane linkage when the unmodified dextran was used and via

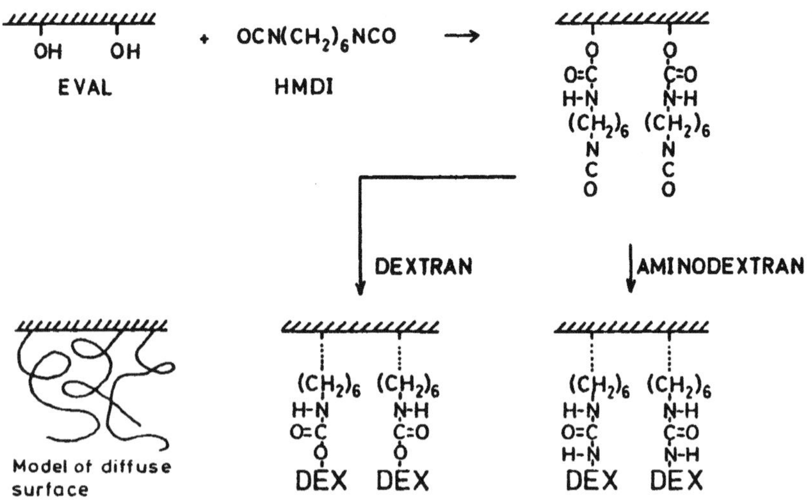

FIGURE 7.1. Reaction scheme for coupling of dextran or aminodextran to produce a diffuse surface. EVAL—ethylene-vinyl alcohol copolymer; HMDT—hexamethylene diisocyanate. (Source: Taniguchi et al., 1982 [2].)

urea linkage when the dextran had amino groups. Urea linkage was formed at a much faster rate than urethane in the absence of catalyst. The surface of the grafted materials was hydrophilic and became slippery when grafting was allowed to proceed to a high extent.

Our group [3] further investigated a grafting reaction of collagen molecules onto the surface of cellulose and PVA films via covalent bonds using the cyanogen bromide (CNBr) activation method [4]. The coupling reaction onto the film pieces activated by CNBr was carried out at 20°C in sodium bicarbonate solution (pH 9.5) containing collagen. The amount of bound proteins was larger for cellulose than for PVA. When p-toluenesulfonyl chloride was used to activate the cellulose film, this activation method was found to be effective in grafting the collagen as well. The cellulose film became brittle and weak after activation with CNBr, but the PVA film did not.

7.2 GRAFT POLYMERIZATION

Unlike the above chemical coupling reactions, which require reactive groups on both the substrate and the polymer chain, graft polymerization of monomers requires generation of active species to initiate polymerization. If polymerization proceeds via the radical mechanism, either free radicals or peroxides should be generated in the surface region of materials to initiate the polymer chain growth of monomers. The graft polymerizations to be described below all proceed via the radical mechanism, and will be classified on the basis of the initiation mode.

7.2.1 Chemical Initiation

Graft polymerization of acrylamide onto polyetherurethane was investigated in aqueous solution using ceric ammonium nitrate as an initiator [5]. It was proposed that complexation of polyurethane substrate (UH) with ceric ion yielded polymer radicals, which initiated graft polymerization.

$$UH + Ce^{4+} \rightleftharpoons [complex] \rightarrow U + Ce^{3+} + H^+ \qquad (7.2)$$

Samal et al. [6] introduced azo groups into the surface region of polymer films possessing hydroxyl groups (PVA, VAECO, and cellulose), followed by graft polymerization of various hydrophilic monomers. The azo groups were introduced either through urethanation of the surface hydroxyl groups with hexamethylene diisocyanate or direct esterification with 4,4′-azobis-4-cyanovaloyl chloride (AIVC), which is capable of initiating radical polymerization.

$$-OH + OCN-(CH_2)_6-NCO \rightarrow -OCONH-(CH_2)_6-NCO \quad (7.3)$$

$$-OH + ClCO(CH_2)_2-\overset{\displaystyle CH_3}{\underset{\displaystyle CH_3}{\overset{\displaystyle /}{\underset{\displaystyle \backslash}{C}}}}-N=N-\overset{\displaystyle CH_3}{\underset{\displaystyle CH_3}{\overset{\displaystyle /}{\underset{\displaystyle \backslash}{C}}}}-(CH_2)_2 \quad COCl \rightarrow$$

$$-OCO(CH_2)_2-\overset{\displaystyle CH_3}{\underset{\displaystyle CH_3}{\overset{\displaystyle /}{\underset{\displaystyle \backslash}{C}}}}-N=N-\overset{\displaystyle CH_3}{\underset{\displaystyle CH_3}{\overset{\displaystyle /}{\underset{\displaystyle \backslash}{C}}}}-(CH_2)_2 \quad COCl + HCl \quad (7.4)$$

Ampules containing the AIVC-bound film and the monomer (acrylic acid, methyl methacrylate, N-vinyl pyrrolidone, etc.) were degassed, sealed, and kept at 50°C to induce graft polymerization.

Sacak et al.[7] investigated the graft copolymerization of methacrylic acid onto PET fibers initiated by radicals formed from thermal decomposition of benzoyl peroxide. The graft yield increased up to 85°C, and then decreased with a further increase in temperature. The graft yield was dependent on the ratio of the water/solvent mixtures. Osipenko and Martinovicz [8] also grafted acrylic acid onto PET films and fibers using benzoyl peroxide. It was found that the preswelling of PET in dichloroethane led to changes in its sorption-diffusion properties and favored an increase in the degree of grafting, and the addition of the Fe(II), Ni(II), and Cu(II) salts to monomer solution decreased homopolymer yield.

7.2.2 Ozone Initiation

The use of ozone for surface oxidation and surface graft polymerization is a rather classic method of initiation. Graft polymerization of monomers onto an ozonized polymer substrate by the radical mechanism has been described by a number of authors. For instance, Landler and Lebel [9] made a detailed study of graft polymerization onto PVC. The reaction of PVC with ozone and the thermal behavior of the ozonized polymer showed that PVC easily underwent ozonization under formation of thermally labile groups. Ozonization and the subsequent decomposition formed peroxide groups that were not accompanied by considerable degradation of the PVC polymer chain. Landler and Lebel studied graft polymerization of various ethylene monomers onto the PVC surface previously subjected to ozonization. The ozone pretreatment method was also utilized by Korshak et al. [10], who could graft polymerize different monomers on the surface of films, fibers, and fabrics using this method. The types of film and fiber used were polyamide, polyester, fluorinated plastics, and natural fiber (wool and silk), while the vinyl monomers used were styrene, methyl methacrylate, methyl acrylate, acrylonitrile, acrylic and methacrylic acids, and 2-methyl-5-vinyl-pyridine.

Ozone-induced graft polymerization was also carried out onto PU film [11]. The peroxide production at the surface region of the film could be readily controlled by the ozone concentration and exposure time.

7.2.3 Electron Beam Irradiation

Electron beams have been used for surface grafting by relatively few investigators, compared to other high-energy radiations. Lavielle and Schultz [12] reported graft polymerization of acrylic acid onto PE using electron beam radiation. A PE film was first peroxidized by irradiation with an electron beam (0.7 − 1.2 Mrad) in the presence of air. Dilute aqueous acrylic acid solution was then added by spraying to induce grafting on the peroxidized PE. Lavielle and Schultz also studied thermodynamic properties of the grafted PE film [13].

We also performed surface graft polymerization of acrylamide onto a PE film pre-irradiated with an electron beam [14]. First, a PE film was exposed to the electron beam in air to a dose of 10 Mrad, stored in a refrigerator, and subsequently graft polymerized in aqueous monomer solution in the presence of ferrous ions at 15°C. Since the graft density was too low to be determined gravimetrically, the occurrence of grafting was confirmed both by the contact angle measurement and XPS analysis. Figure 7.2 shows the result of the contact angle measurement for the surface-grafted PE films.

7.2.4 Gamma-Ray Irradiation

Ionizing radiation such as gamma rays produces numerous radicals and ionized species deep in the bulk phase of the polymer, because such a high-energy radiation can penetrate through polymeric materials to a considerable depth.

Gamma-ray-induced graft polymerization can be carried out by either of the following methods: (1) simultaneous irradiation and grafting through formed free radicals, (2) grafting through peroxide groups introduced by pre-irradiation, and (3) grafting initiated by trapped radicals formed by pre-irradiation.

In 1960, Sakurada [15] stressed the potentiality of the grafting process for fiber modification. Nevertheless, so far, studies on surface graft polymerization with the use of gamma rays are very few, compared with studies of other high-energy sources. The following are examples of such studies.

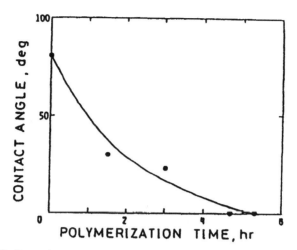

FIGURE 7.2. Dependence of contact angle on the polymerization time for HDPE pre-irradiated in air with electron beams to 10 Mrad (polymerization at 15 °C and 25 wt% of AAm in the presence of Fe^{2+} of 1×10^{-4} mol·1^{-1}).

Most graft polymerizations with the use of gamma rays have been applied to synthesis of hydrogel onto polymer surfaces, except for a few studies such as the one by Rosiak et al. [16], who performed cross-linking of poly(acrylamide) by irradiation. A hydrogel may be defined as a polymeric material that exhibits the ability to swell in water and retain a significant fraction of water within its structure, but which will not dissolve in water. The idea of using hydrophilic polymer gels as medical materials was proposed for the first time by Wichterle about thirty years ago [17]. Chapiro [18] studied hydrogel formation by radiation grafting to improve the thromboresistance of polymeric materials. However, one of the chief objectives of surface modification by gamma-ray irradiation is usually to synthesize hydrogel-like materials.

Hoffman et al. [19] extensively studied surface graft polymerization using a mixture of HEMA and ethyl methacrylate. Figure 7.3 shows effects of the monomer ratio on the graft level at different radiation doses. It is apparent that the graft reaction takes place faster in the beginning if HEMA is present at a high concentration in the monomer mixture. The graft layer thickness estimated for HEMA/PE grafts is shown in Table 7.1. The grafted polymer was stained by reacting the hydroxyl group of HEMA with cinnamyl chloride to attach a double-bond molecule to the HEMA molecule, followed by reacting the double bond with osmium tetraoxide. As can be seen from Table 7.1, the graft layer is thicker than several tens of μm. Hoffman also prepared such hydrophilic polymeric biomaterials either utilizing gamma rays or glow discharge methods [20]. Lawler and Charlesby [21] studied the grafting of acrylic acid onto PE surfaces in

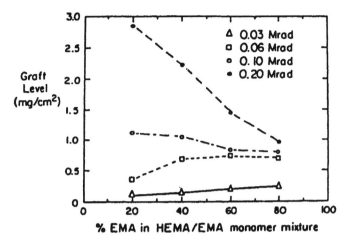

FIGURE 7.3. Effect of monomer composition on HEMA/EMA graft copolymerization on PE (solvent = 90.6% ethanol, 9.4% H_2O). ©1983. Reprinted from *Radiat. Phys. Chem.*, p. 275, Hoffman, A. S. et al. [19] with permission from Pergamon Press, U.K.

aqueous solutions using gamma radiation, and found that the PE-grafted surface was tightly hydrogen bonded and deformed, and made fissures to allow fresh monomers to penetrate into PE. Further grafting soon became diffusion-controlled and the degree of penetration of acrylic acid into PE depended on time.

To improve blood compatibility, surface modification of a hydrophobic polymer was achieved by radiation-induced graft polymerization by our group [22]. A PE film placed in a glass tube was sealed under an atmosphere of dry air and then irradiated at room temperature with gamma rays from a Co^{60} source. Following irradiation, the film was put into aqueous acrylamide solution in an ampule and sealed after degassing. To effect graft polymerization, the mixture was heated to 50°C, or an adequate amount of

TABLE 7.1. Estimated Graft Layer Thickness for HEMA/PE Grafts.

Extent of Grafting (mg/cm²)	Approximate Graft* Layer Thickness (μm)
1.58	15
2.11	20
2.51	30
2.72	45

*Estimated from light micrographs of HEMA/PE graft cross sections stained by sequential reactions with cinnamyl chloride and osmium tetraoxide. ©1983. Reprinted from Hoffman, A. S. et al. [19], *Radiat. Phys. Chem.*, p. 272, with permission from Pergamon Press Ltd., U.K.

ferrous sulfate (FeSO$_4$ $10^{-4} - 10^{-3}$ M) was added to the monomer solution before sealing for redox graft polymerization.

The PE surface was also grafted with acrylamide by Postnikov et al. [23]. The PE pre-irradiated by gamma rays from a Co^{60} source (1 – 20 Mrad) in air was brought into contact with deaerated monomer -Fe^{2+} aqueous solution. They studied the effect of Fe^{2+} concentration in a redox system on graft polymerization at 40°C.

Jansen [24] reported the "preswelling technique" for the modification of PU surfaces. A PU tube was preswelled with a monomer solutions such as HEMA and acrylamide or filled with a monomer stream, and subsequently a gamma-irradiated to induce graft polymerization. Jansen and Ellinghorst [25] studied grafting of hydrophilic or reactive monomers (HEMA, 2,3-epoxypropyl methacrylate, 2,3-dihydroxypropyl methacrylate, and acrylamide) by gamma radiation. They found that the water uptake and diffusion of the grafted and subsequently chemically modified film increased with the grafting yield. The degree of hydrophilicity, especially, of HEMA-grafted film strongly depended on grafting conditions. For some grafted samples with high surface hydrophilicity, very low interfacial free energies between the surface and water were found.

The grafting by preirradiation technique using gamma rays was reported by Hegazy [26], who studied the influence of grafting parameters such as pre-irradiation dose, monomer concentration, and grafting temperature on the grafting yield for a combination of a hydrophilic monomer and fluorinated polymer films.

Graft polymerization induced by gamma radiation was applied for preparing a surgical device with water-swellable sleeves to facilitate an access to the body cavity [27]. An extruded tube (the sleeve material) of EVA having an internal longitudinal channel and holes, was placed in a glass vessel containing an aqueous solution of acrylic acid and ferrous ion. The mixture was evacuated and sealed, followed by irradiation with gamma rays from a Co^{60} source to a total dose of 1 Mrad.

7.2.5 Plasma Treatment

7.2.5.1 GLOW DISCHARGE

The use of low-temperature plasmas has been extensively studied to modify surface properties of polymer substrates, as was described in Chapter 5. The major methods are either plasma polymerization, by which a cross-linked thin polymeric layer is deposited on the substrate surface, or plasma treatment, by which intensive oxidation is brought about on the surface region of the substrate [28]. However, it is also possible to modify the polymer surface by graft polymerization utilizing free radicals or

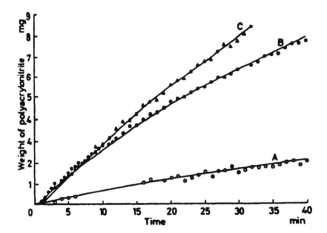

FIGURE 7.4. Graft polymerization of acrylonitrile onto high-density polyethylene at 80 °C. A—untreated polyethylene (71.5 mg); B—polyethylene (103 mg) submitted to the discharge at 77 °K; C same as B, but the specimen (130 mg) was subsequently allowed to oxidize in air at 20 °C for 17 hours. Zero time in these experiments was difficult to determine because extensive bumping often occurred when the dilatometers were immersed in the thermostat at 80 °C. (Source: Bamford and Ward, 1961 [29].)

peroxides generated by the plasma treatment similar to irradiation with high-energy radiations.

As already described in Chapter 5, Bamford and Ward disclosed that graft polymerization of vinyl monomers could take place on a polymer surface treated with a Tesla coil [29]. In their experiment, high-frequency discharge was applied to polyethylene powder in dilatometer containing hydrogen at a pressure of 0.1 mm Hg. The dilatometer was then filled with degassed acrylonitrile, followed by polymerization at 80°C. Graft levels of acrylonitrile onto the PE powder were up to 50 wt%. The results of some typical dilatometer experiments are shown in Figure 7.4. The radical concentration formed on the surface of the crystalline substrate by the Tesla coil treatment was on the order of a unimolecular layer [30]. The variation of radical concentration with time at different temperatures is shown in Figure 7.5.

Other workers also investigated graft polymerization onto polymer surfaces exposed to glow-discharge plasma of inert gas. The free radicals formed as a result of the plasma treatment were utilized for the subsequent propagation reaction of monomers in contact with the plasma-treated surface. Fales et al. [31] exposed textile materials to argon plasma to introduce free radicals. Other research groups [32–34] also used plasma treatment to yield radicals for surface modification of polymers.

A different graft polymerization method was also reported [34]. First, a

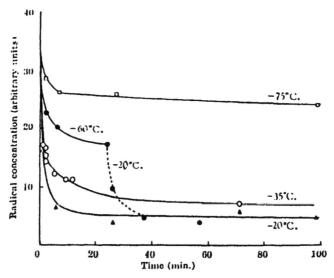

FIGURE 7.5. Decay of surface radicals on methacrylic acid crystals at various temperatures. Scale of radical concentration is arbitrary, but 30 units = 10^{-3} mol l^{-1}, approximately. Different symbols are used to distinguish between different experiments. (Source: Bamford et al. [30].)

polymer was exposed to plasma, followed by contacting it with air prior to the polymerization procedure. The peroxide formed after the plasma exposure was utilized to initiate graft polymerization onto the plasma-treated polymer. Site dependence of peroxide formation by plasma treatment has been studied by Piao et al. using PET fabrics [36]. To estimate the site distribution, a new parameter $A(x) = Pm(x)/P(\text{max})$ was introduced, where $Pm(x)$ and $Pm(\text{max})$ represent the density of peroxides at site x and the maximum density of peroxides generated on the PET surface, respectively. If parameter $A(x)$ is close to 1, the distribution of peroxides would be homogeneous. The site distribution of peroxides on the plasma-treated PET fabric is shown in Figure 7.6 and Table 7.2. The results revealed that the distribution was influenced by the relative positions of the electrodes and the direction of the plasma flow.

An advantage of the plasma-induced graft polymerization method over the gamma ray radiation method is that the location of grafting is definitely limited to the surface region of plasma-treated polymeric materials, independent of the polymerization conditions. To give evidence of this, an optical microscopic photograph of the cross section of PE film grafted with acrylamide is shown in Figure 7.7. The grafting condition employed was as follows. The PE film exposed to air after plasma treatment was immersed in 10 wt% aqueous solution of acrylamide in a glass ampule. After rigorous degassing, the ampule was sealed and kept at 50°C for 1 hour. When various

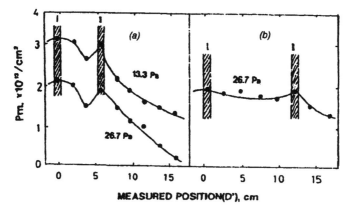

FIGURE 7.6. Space distribution of peroxide yields (*Pm*) along the center line in the reaction tube. Poly(ethylene terephthalate) fabrics were placed at the distance D' from electrode I and exposed to plasma for 60 s. Electrode separations were 6 cm (a) and 12 cm (b) [36].

gases (argon, oxygen, nitrogen, and hydrogen) were used for plasma treatment, the results shown in Figure 7.8 were obtained for graft polymerization onto the PE films. It is interesting to note that graft density varies, passing through a maximum, with the plasma exposure time, and, in addition, that nitrogen plasma does not lead to any appreciable graft polymerization within the range of exposure time studied.

Hsieh and Wu [37] investigated surface grafting of acrylic acid on argon glow-discharged PET films. Although the wettability decay was observed for the PET films exposed to argon plasma during the initial few days after the treatment, acrylic acid-grafted PET surfaces maintained their wettability for both the liquid-phase and the vapor-phase graft polymerizations. The residual reactivity of the glow-discharged surfaces was captured by

TABLE 7.2. Effects of Electrode Distance and Gas Pressure
on the Parameter A Calculated from the Peroxide Yields
on the PET Fabrics Treated with Ar Plasma for 60 s.

$A(x)$	Gas Pressure (P) (Pa)	Electrode Distance (D) (cm)		
		6	9	12
$A(D' = 14$ cm)	26.7	0.31	0.92	0.89
A(av)	26.7	0.59	0.75	0.86
A(min)	26.7		0.63	
	53.3		0.63	
A(min)	13.3		0.81	
	26.7		0.80	

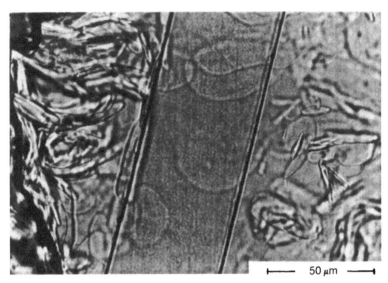

FIGURE 7.7. Optical microscopic cross section of the PE film grafted with PAAm after hydrolysis and staining (graft density was 90 $\mu g \cdot cm^{-2}$). (Source: Suzuki et al., 1986 [35].)

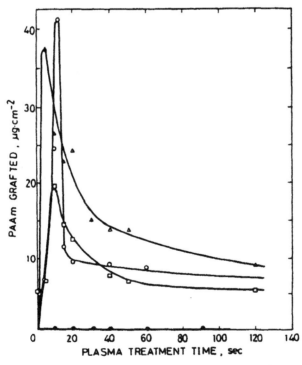

FIGURE 7.8. Dependence of the graft density on plasma exposure time for the PE film exposed to different plasmas of 11.5 W followed by graft copolymerization: (○) Ar; (△) O_2; (□) H_2; (●) N_2. (Source: Suzuki et al., 1986 [35].)

exposing the pretreated PET film to acrylic acid in the liquid and vapor phases. They also examined the surface morphology of the grafted surfaces by scanning electron microscopy.

7.2.5.2 CORONA DISCHARGE

Iwata et al. [38] oxidized a PE film surface with plasma generated by corona discharge and subsequently graft-polymerized acrylamide onto the PE surface. The maximum density of peroxides (2.3×10^{-9} mol cm^{-2}) was yielded by corona treatment at a voltage of 15 kV. The decomposition rate of peroxides and the dependence of the graft density on the storage period of the corona-treated PE film showed that there were several types of peroxides, among which the most labile one was mainly responsible for the initiation of graft polymerization. When the corona-treated film was brought into contact with a deaerated aqueous solution of acrylamide, the graft density increased with the treatment time, but then decreased after passing through a maximum. Optical microscopy of the cross section of the grafted film revealed that graft polymerization was limited to a very thin surface region.

Noh et al. [39] reported that the hydroperoxide formed on a polypropylene film and filament upon corona treatment was directly proportional to the corona treatment time when analyzed by iodometry and IR spectroscopy.

7.2.6 Photoirradiation

Photoinduced radical graft polymerization has been extensively studied using UV light. Earlier studies by Oster et al. [40−42] revealed that UV light exposure led successfully to cross-linking of polyisobutylene, PP, and PMMA in the presence of sensitizers, and allowed photo-graft polymerization of monomers onto polymers blended with photosensitizers. One of the objectives of their work was to find substitutes for silver bromide, currently still used in photographic technologies. Charlesby et al. [43] quantitatively analyzed the cross-linking results of PE by UV radiation, reported by Oster et al. A more detailed study was undertaken by Takakura et al. [44], who showed that a PVA film could be cross-linked by UV irradiation in the presence of sodium benzoate as a sensitizer, always accompanied by photolysis of sodium benzoate used as a sensitizer. The gel content increased with irradiation time and approached a limiting value, depending on the initial sensitizer concentration. Bellobono et al. [45] performed photochemical grafting of acrylated azo dyes such as 4-(N-ethyl, N-2-acryloxyethyl) amino, 4′-nitro, azobenzene onto polymeric materials. The substrate polymers used were woven fabrics of nylon 6 and PP. They chose this model system in order to understand the fundamental mechanisms of the grafting process.

The use of UV irradiation appears to be an excellent method because of its simplicity and the cleanliness of the treatment. Tazuke et al. [46] stressed the suitability of the photochemical method for surface grafting of polymers for the following reasons:

(1) Photochemically produced triplet states of carbonyl compounds can abstract hydrogen atoms from almost all polymers so that graft polymerization may be initiated.

(2) High concentrations of active species can be produced locally at the interface between the substrate polymer and the monomer solution containing a sensitizer when photoirradiation is applied through the substrate polymer film.

(3) In addition to the simplicity of the procedure, the cost of the energy source is lower for UV radiation than for ionizing radiation.

Many methods are known to chemically modify polymer surfaces through graft polymerization with the use of UV light. They may be carried out either under a wet condition (graft polymerization in a monomer solution) or under a dry condition (vapor phase graft polymerization).

7.2.6.1 PHOTOIRRADIATION WITH PHOTOSENSITIZER

Tazuke and Kimura [47] succeeded in the photografting of monomers onto PP and several other polymer films using benzophenone as sensitizer and 3 kW xenon lamp as the light source (Figure 7.9). Irradiation for 7 min was sufficient to produce a hydrophilic surface on PP. PVC, poly(vinylidene chloride) (PVdC), and cellulose acetate, which were easily attacked by radicals, could be photografted using methanol as a solvent. PE and PP were more stable against radical attack, so that the use of methanol, which reacts with the triplet state of benzophenone (BP^{*3}) to produce radicals, entirely inhibited graft polymerization, and allowed only homopolymerization of acrylamide. This inhibition of grafting indicates that the correct choice of solvent and sensitizer is very important. Acetone is stable against hydrogen abstraction by BP^{*3}, which can, however, abstract hydrogen atoms from polymers to initiate graft polymerization. This is supported by the result that addition of 0.25 M isopropanol to acetone almost inhibits graft polymerization onto polypropylene although homopolymerization is not affected. Therefore, graft polymerization is probably initiated either by hydrogen abstraction by BP^{*3} or by radical attack; the former process alone is applicable to polyolefins.

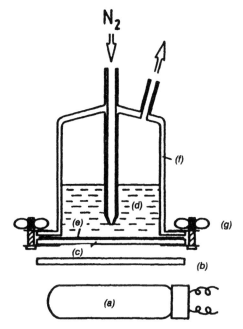

FIGURE 7.9. Apparatus for photochemical surface grafting. (a) 100 W high-pressure mercury lamp; (b) interference filter; (c) Pyrex plate; (d) solution containing monomer and initiator; (e) polymer film; (f) glass vessel; (g) clamp [47].

Photografting of vinyl monomers has also been studied to render poly(dimethyl siloxane) surfaces hydrophilic [48]. The investigators used a chlorine-containing poly(dimethyl siloxane) (C-PDMS) which was prepared by polymerization of chloromethyl-heptamethyl cyclotetrasiloxane and photocuring on a glass plate. The cross-linked C-PDMS was subsequently subjected to a reaction with sodium diethyl dithiocarbamate. The diethyl dithiocarbamated PDMS was then photoirradiated in solutions of hydrophilic vinyl monomers such as HEMA and acrylamide to yield the surface-grafted PDMS. The solvents employed were methanol, acetone, or their mixture for HEMA, and water or a water-methanol mixture for acrylamide.

Graft polymerization of vinyl monomers such as acrylic acid and methacrylic acid on a benzophenone-coated LDPE film in aqueous medium was performed by Ogiwara et al. [49]. The benzophenone-coated LDPE was prepared by immersing the LDPE film in an acetone solution containing 0.3% benzophenone and 1.0% poly(vinyl acetate), followed by drying under reduced pressure. However, it was not easy to graft-polymerize other hydrophilic monomers such as acrylamide to a high level by this photo-initiator-coating method. Therefore, they used a two-step process for the photoinduced grafting [50]. The first step was carried out by the same method described above, and then the grafted sample was coated with benzophenone again to prepare it for the second step, grafting. The first-step LDPE film was placed in a Pyrex glass tube with aqueous acrylamide solution, flushed with nitrogen gas, and then irradiated at 60°C to allow polymerization to proceed. ESR spectra of the methacrylic acid or acrylic acid-grafted LDPE, both coated with benzophenone, were analyzed after UV irradiation, and the following radicals were assumed to be generated on the LDPE surface:

$$
\begin{array}{c}
CH_3 \\
/ \\
\text{\textasciitilde\textasciitilde\textasciitilde\textasciitilde } CH_2 - CN\cdot \qquad\qquad I \\
\backslash \\
COOH
\end{array}
$$

$$
\begin{array}{c}
H \\
/ \\
\text{\textasciitilde\textasciitilde\textasciitilde\textasciitilde } CH_2 - CN\cdot \qquad\qquad II \\
\backslash \\
COOH
\end{array}
$$

The preirradiated acrylic acid-grafted LDPE was found to initiate grafting of acrylic acid. The irradiated sample probably contains primarily radical II, because the ketyl radical based on benzophenone is thermally unstable and cannot exist at room temperature. Therefore, radical II has most likely initiated further grafting with the acrylamide.

7.2.6.2 PHOTOIRRADIATION WITHOUT PHOTOSENSITIZER

Most of the photoinduced graft polymerizations have been carried out using photosensitizers, as described above. However, it is well known that UV irradiation results in buildup of hydroperoxides in many polymers, regardless of the presence of photosensitizers, followed by polymer degradation [51]. Although the density of the formed peroxides may be low, these peroxides are expected to function as efficient initiators for surface grafting, similar to those obtained by irradiation with high-energy radiation.

We performed photoinduced graft polymerization onto the film surfaces of PP, PE, and EVA without utilizing any photosensitizer and photoinitiator [52]. Films were pre-irradiated with UV from a high-pressure mercury lamp in air, and then immersed in a deaerated aqueous solution of acrylamide at 50°C. Graft polymerization took place in high yields with the increasing UV pre-irradiation time, and the density of the generated peroxides on the film surface was proportional to the UV exposure. Graft polymerization took place effectively also onto pre-irradiated EVA films having various vinyl acetate (VA) contents, which influenced the density of grafted PAAm, as Figure 7.10 shows. The EVA without VA component, that is, PE, was not grafted at all even under the prolonged irradiation (10 hours). Since PP is very vulnerable to photodegradation because of the tertiary carbon atoms present along the polymer chain, much effort has been directed to preventing PP from degradation and to stabilizing it during the UV exposure.

Among the synthetic polymers classified by Somersall and Guillet [53], a group that absorbs and emits light through isolated chromophore impurities which are located in-chain or as end-chain groups, theoretically should not absorb UV light of wavelengths longer than 290 nm. This value is practically the lower limit of terrestrial sunlight. As pure polyolefins do not contain any chromophores, they should not be affected by exposure to UV radiation with longer than 200 nm, as well as radiation of natural sunlight. Nevertheless, PP degrades via photooxidation. This is probably because certain impurities such as catalyst residues, aromatic compounds

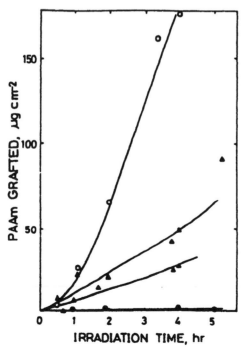

FIGURE 7.10. Effect of UV irradiation time on graft polymerization of acrylamide onto polyethylene and ethylene-vinyl acetate copolymer films having various vinyl acetate contents: (●) VA 0% (PE); (○) VA 28%; (△) VA 19%; (▲) VA 14%.

(acting as a chromophore), unsaturated carbonyl groups, or hydroperoxides (initiating dehydrogenation from the main chain of the polymer) may have been incorporated to commercial polyolefins during the process of polymerization.

7.2.6.3 WITHOUT DEGASSING

In all of the methods mentioned above, exclusion of oxygen from the monomer solution is essential for surface graft polymerization. This is the greatest obstacle for industrial applications of the methods, as oxygen exclusion from the polymerization system is a very time-consuming and expensive process. Further, it usually involves purging with an inert gas or repeated cycles of freezing and thawing. It is, therefore, desirable to omit the oxygen removal process, especially for surface graft polymerization on a large scale.

We [54] found a new method that omits the oxygen removal procedure from polymerization mixtures when graft polymerization of acrylamide is carried out onto the surface of PET film. The method involves simultaneous UV irradiation without utilizing a photosensitizer, but the monomer mixture contains an appropriate quantity of periodate ($NaIO_4$), which consumes oxygen during the irradiation. The change in the oxygen concentration during UV irradiation for 10 wt% acrylamide solution containing different concentrations of $NaIO_4$ is given in Figure 7.11, together with the contact angle of PET films grafted under the same experimental conditions. The UV irradiation was performed with a high-pressure mercury lamp (40 W) for 90 min at 35°C. When the concentration of $NaIO_4$ was below 1×10^{-4} M or above 1×10^{-2} M, polymerization did not take place, but proceeded to a significant extent when the $NaIO_4$ concentration was in a range between 3×10^{-4} and 5×10^{-3} M. Figure 7.12 shows the decrease in the oxygen concentration in the monomer mixture during the course of graft polymerization.

The graft polymerization method without oxygen removal process was also performed for PMMA [55]. In this work, riboflavin was used instead of periodate to effect graft polymerization of acrylamide onto a PMMA plate with various treatments for introducing peroxides onto the PMMA surface, such as UV pre-irradiation, corona discharge, glow discharge, and ozone exposure. When the graft polymerization was performed by near-UV light irradiation to decompose the peroxides in the presence of riboflavin, the graft density was $20-25$ μg/cm² with the contact angle around 20° for the PMMA pre-treated with corona discharge, glow discharge, and UV irradiation. However, graft polymerization practically did not take place onto the PMMA plate when pre-treated with ozone exposure. When graft polymerization onto the pre-treated PMMA was performed at 50°C after

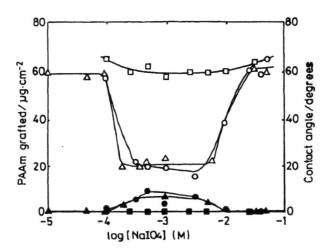

FIGURE 7.11. Effect of AAm concentration on AAm graft polymerization onto PET films without degassing by UV irradiation (35 °C, 90 min). [AAm] (wt%): (□, ■) 5; (△, ▲) 10; (○, ●) 15; open marks—contact angle; closed marks—amount of PAAm grafted [54].

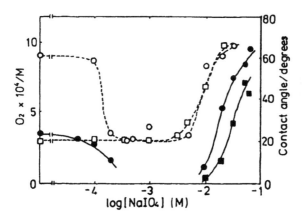

FIGURE 7.12. Change of O_2 concentration of AAm solution containing $NaIO_4$ of different concentrations during UV irradiation (10 wt% AAm, 35 °C, 90 min): (○, ●) without degassing; (□, ■) under degassing; solid lines: O_2 concentration; broken lines: contact angle [54].

degassing but without UV irradiation and riboflavin, the graft density was always less than 10 μg/cm^2, independent of the pre-treatment methods.

7.2.6.4 VAPOR-PHASE POLYMERIZATION

Photoinduced graft polymerization described above often produces a large amount of homopolymers in the polymerization mixture. This problem may be overcome by carrying out the graft polymerization in the vapor phase of the monomer. Kachan et al. [56] irradiated polyamide films directly with UV light of 234 nm *in vacuo* at room temperature and subsequently exposed them to vapor of acrylonitrile, vinyl acetate, vinylpyridine, and isoprene. The absence of a homopolymer in vapor-phase grafted polymerization made it possible to take its kinetics continuously during the reaction in adsorption vessels with the aid of MacBen scales.

Ogiwara et al. [49,57,58] also studied vapor-phase graft polymerization in conjunction with UV irradiation. Prior to graft polymerization, polymer films were pre-treated by dipping them into acetone mixtures containing poly(vinyl acetate) and a photoinitiator such as benzophenone, anthraquinone, and benzoyl peroxide. The dried films were then placed in a Pyrex glass tube with the monomer gas. The system was exposed to UV from a high-pressure mercury lamp at 10°C, as shown schematically in Figure 7.13. As the film contained a high concentration of photoinitiator at the surface, a high graft density was obtained.

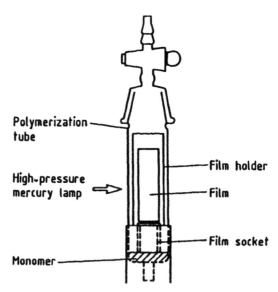

Polymerization tube

High-pressure mercury lamp

Film holder

Film

Film socket

Monomer

FIGURE 7.13. Apparatus for photochemical vapor phase surface grafting (after Ogiwara et al. [49]).

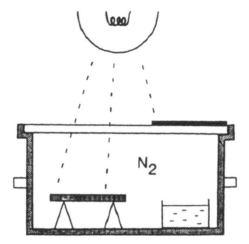

FIGURE 7.14. Reactor for surface grafting. The LDPE sample is irradiated with an UV lamp. Monomer and initiator are evaporated from a beaker which also contains a solvent. (Source: Allmer, K., A. Hult and B. Rånby. 1989. *J.P.S. Polym. C.*, 27:1641.)

Rånby et al. [59] studied surface modification of low-density polyethylene and polypropylene by photoinduced graft polymerization of acrylic acid. In their grafting technique, the polymer specimen to be grafted was placed in a small reactor, together with a beaker containing a solution of 2 M acrylic acid and 0.2 M benzophenone in acetone. The apparatus employed in their vapor phase graft polymerization system is shown in Figure 7.14. To avoid the interference of radical polymerization by oxygen in air, the reactor was first purged with nitrogen for 10 min, sealed, heated to 60°C in a water bath, and irradiated with UV through a quartz window. Since both the monomer and the initiator were supplied in the vapor phase, freely standing and UV-transparent specimens were grafted on both sides. Grafting was affected by the combination of solvent/carrier used and the type of polymer. Acetone was able to promote direct grafting to the surface. Polystyrene was easier to graft than other polymers, presumably because it contained easily abstractable tertiary hydrogen.

A system for processing continuous photograft polymerization of acrylamide and acrylic acid onto the surface of HDPE tape film has been developed by Rånby's group, as shown in Figure 7.15 [60,61].

Goldblatt et al. [62] modified a PET film surface using a method similar to that of Rånby et al. to study adhesion to vacuum-deposited copper [63].

7.2.7 Far-UV Irradiation

In contrast to radiation of wavelengths longer than 200 nm, far-ultraviolet (far-UV) offers several attractive features for the modification of polymer surfaces. Lazare et al. [64] pointed out that far-UV radiation (180–200 nm) has a small penetration depth in organic solids because of their high absorption and thus should be much more adapted to surface reactions, especially to thin film modifications. Indeed, most of the studies for surface modification of polymers with far-UV radiation have been performed at wavelengths longer than 200 nm, which is only weakly absorbed by oxygen in air and does not require a vacuum system [65]. Optical elements and vessels made of quartz permit transmittance of this radiation. The energy of a far-UV photon (>6 eV or 138 kcal/mol) exceeds the energy of most covalent bonds in polymer chains and, therefore, has a high possibility of photochemical bond breaking. Surface analysis of PET films exposed to far-UV radiation was performed by both XPS and iodometry. Typical results are shown in Figure 7.16. XPS spectra of the oxidized PET film showed a decrease in O/C ratio, when the film was dipped into water for 1 hour prior to XPS analysis. The result indicates that most, if not all, of the oxidized substances introduced on the PET surface transferred to the liquid phase, probably because a highly oxidized, nonvolatile layer was removed from the extensively photooxidized surface.

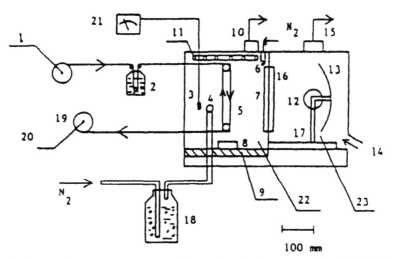

FIGURE 7.15. A sketch of a surface photografting device: (1) tape film feed roll; (2) presoaking solution; (3) thermocouple (screened from UV lamp); (4) vapor inlet of monomer and initiator; (5) running rolls; (6) nitrogen inlet; (7) quartz window; (8) container of solid monomer; (9) electric heater; (10) exhaust outlet; (11) cooling water pipe; (12) UV lamp; (13) parabolic reflector (of aluminium); (14) air inlet; (15) air outlet (ventilation); (16) screen; (17) lamp support; (18) monomer and initiator bubble solution; (19) taking off roll; (20) driving motor; (21) temperature indicator; (22) reaction chamber; (23) lamp box. (Source: Yao and Rånby, 1990, p. 1647 [61].)

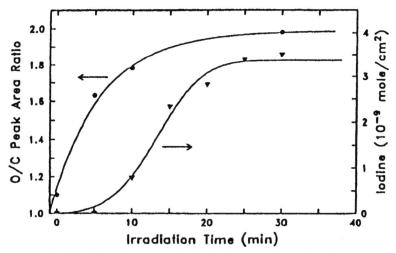

FIGURE 7.16. The effect of exposure of poly(ethylene terephthalate) at 185 nm in air. (●) XPS O/C area ratio as a function of irradiation time. (▼) Amount of iodine produced per cm² of irradiated surface as a function of irradiation time [64].

7.3 REFERENCES

1. Hommel, H., A. P. Legrand and P. Tougne. 1984. "Influence of the Grafting Ratio on the Conformations of Poly(ethylene oxide) Chains Grafted on Silica," *Macromolecules,* 17:1578–1581.

2. Taniguchi, M., R. K. Samal, M. Suzuki, H. Iwata and Y. Ikada. 1982. *Grafting Reaction of Dextran onto Polymer Surface,* Ame. Chem. Soc. Symp. Ser., No. 187, pp. 217–232.

3. Tabata, Y., S. V. Lonikar, F. Horii and Y. Ikada. 1986. "Immobilization of Collagen onto Polymer Surfaces Having Hydroxyl Groups," *Biomaterials,* 7:234–238.

4. March, S. C., I. Parikh and P. Cuatrecasas. 1974. *Anal. Biochem.,* 60:149.

5. Giu. K.-Y., F.-L. Yang, Z.-X. Huang, X.-J. Feng and X.-D. Feng. 1982. "Grafting Polymerization of Acrylamide on Polyether Urethane Film Initiated by Ceric Salt," *Gaofenzi Tongxun* (in Chinese), (2):81–86.

6. Samal, R. K., H. Iwata and Y. Ikada. 1983. In *Physicochemical Aspects of Polymer Surfaces, Vol. 2,* K. L. Mittal, ed., Plenum, pp. 801–815.

7. Sacak, M., F. Sertkaya and M. Talu. 1992. "Grafting of Poly(ethylene terephthalate) Fibers with Methacrylic Acid Using Benzoyl Peroxide," *J. Appl. Polym. Sci.,* 44:1737–1742.

8. Osipenko, I. F. and V. I. Martinovicz. 1990. "Grafting of the Acrylic Acid on Poly(ethylene terephthalate)," *J. Appl. Polym. Sci.,* 39:935–942.

9. Landler, Y. and P. Lebel. 1960. "Greffage sur Polychlorure de Vinyle par Pre'-Ozonisation," *J. Polym. Sci.,* 28:477–489.

10. Korshak, V. V., K. K. Mozgova and M. A. Shkolina. 1964. "Grafting as a Method of Surface Modification of Heterochain Polymers," *J. Polym. Sci., Part C,* 32(4):753–764.

11. Fujimoto, K., Y. Takebayashi, H. Inoue and Y. Ikada. 1993. "Ozone-Induced Graft Polymerization onto Polymer Surface," *J. Polym. Sci., Part A, Polym. Chem.,* 31:1035–1043.

12. Lavielle, L. and J. Schultz. 1985. "Surface Properties of Graft Polyethylene in Contact with Water I. Orientation Phenomena," *J. Colloid Interface Sci.,* 106:438–445.

13. Lavielle, L., J. Shultz and A. Sanfeld. 1985. "Surface Properties of Graft Polyethylene in Contact with Water II. Thermodynamic Aspects," *J. Colloid Interface Sci.,* 106:446–451.

14. Ikada, Y. and M. Suzuki. 1982. "Hyomenkaishitu no Tameno Houshasen Gurafutojugou (Radiation-Induced Graft Polymerization for Surface Modification)," *Kasen Koenkai Prep.* (in Japanese), 39:59–69.

15. Sakurada, I. 1960. *Kagaku* (in Japanese) 15:828–832.

16. Rosiak, J. et al. 1983. *Radiat. Phys. Chem.,* 22:917.

17. Wichterle, O. and D. Lim. 1960. "Hydrophilic Gels for Biological Use," *Nature,* 185:117–118.

18. Chapiro, A. 1983. "Radiation Grafting of Hydrogels to Improve the Thrombo-Resistance of Polymers," *Eur. Polym. J.,* 19:859–861.

19. Hoffman, A. S., D. Cohn, S. R. Hanson, L. A. Harker, T. A. Horbett, B. K. Ratner and L. O. Reynolds. 1983. "Application of Radiation-Grafted Hydrogels as Blood-Contacting Biomaterials", *Radiat. Phys. Chem.,* 22:267–283.

20. Hoffman, A. S. 1984. "Ionizing Radiation and Gas Plasma (or Glow) Discharge Treatments for Preparation of Novel Polymeric Biomaterials," *Adv. in Polym. Sci.,* 57:141–157.

21. Lawler, J. P. and A. Charlesby. 1980. "Grafting of Acrylic Acid onto Polyethylene Using Radiation as Initiator," *Radiat. Phys. Chem.,* 15:595–602.

22. Suzuki, M., Y. Tamada, H. Iwata and Y. Ikada. 1983. In *Physicochemical Aspects of Polymer Surfaces, Vol. 2,* K. L. Mittal, ed., Plenum, pp. 923–941.

23. Postnikov, V. A., N. J. Lukin, B. V. Maslow and N. A. Plate. 1980. "The Simple Preparative Synthesis of Graft Copolymers of Polyethylene-acrylamide," *Polym. Bull.,* 3:75–81.

24. Jansen, B. 1983. "Radiation Induced Modification of Polyetherurethane Tubes with HEMA and Acrylamide," *Polym. Sci. Tech.,* 23:287–295.

25. Jansen, B. and G. Ellinghorst. 1984. "Modification of Polyetherurethane for Biomedical Application by Radiation Induced Grafting. II. Water Sorption, Surface Properties, and Protein Adsorption of Grafted Films," *J. Biomed. Mater. Res.,* 18:655–669.

26. Hegazy, E.-S. A. 1984. "Preirradiation Grafting of *N*-Vinyl-2-pyrrolidone onto Poly(tetrafluoroethylene) and Poly(tetrafluoroethylene-hexafluoropropylene) Films," *J. Polym. Sci., Polym. Chem. Ed.,* 22:493–502.

27. Miller, R. A. U.K. patent application GB 2179258, 1987.

28. Yasuda, H. K. 1984. "Plasma Polymerization and Plasma Treatment," *J. Appl. Polym. Sci., Appl. Polym. Symp.,* 38.

29. Bamford, C. H. and J. C. Ward. 1961. "The Effect of the High-Frequency Discharge on the Surface of Solids. I. The Production of Surface Radicals on Polymers," *Polymer,* 2:277–293.

30. Bamford, C. H., A. D. Jenkins and J. C. Ward. 1960. "The Tesla-Coil Method for Producing Free Radicals from Solids," *Nature* (London), 186(May 28):712.

31. Fales, J. D., A. Bradley and R. E. Howe. 1976. "Surface Grafting of Textile Materials," *Vac. Technol.* (March):53−56.

32. Wertheimer, M. R. and H. P. Schreiber. 1981. "Surface Property Modification of Aromatic Polyamides by Microwave Plasmas," *J. Appl. Polym. Sci.*, 26:2087−2096.

33. Hatada, K., H. Kobayashi, Y. Masuda and S. Kitano. 1981. "The Graft Polymerization of Fluoroalkylacrylate onto Polyester Fiber Using Cold Plasma and the Optical Properties of the Fiber," *Kobunshi Ronbunshu* (Japan), 38:615−621.

34. Simionescu, C. I., F. Dene's, M. M. Macoveanu and I. Negulescu. 1984. "Surface Modification and Grafting of Natural and Synthetic Fibers and Fabrics under Cold Plasma Conditions," *Makromol. Chem.*, (Suppl. 8):17−36.

35. Suzuki, M., A. Kishida, H. Iwata and Y. Ikada. 1986. "Graft Copolymerization of Acrylamide onto a Polyethylene Surface Pretreated with a Glow Discharge," *Macromolecules*, 19:1804−1808.

36. Piao, D.-X., Y. Uyama and Y. Ikada. 1991. "Space Distribution of Peroxides Generated on Polymer Surfaces by Plasma Exposure," *Kobunshi Ronbunshu*, 48:535−539.

37. Hsieh, Y.-L. and M. Wu. 1991. "Residual Reactivity for Surface Grafting of Acrylic Acid on Argon Glow-Discharged Poly(ethylene terephthalate) (PET) Films," *J. Appl. Polym. Sci.*, 43:2067−2082.

38. Iwata, H., A. Kishida, M. Suzuki, Y. Hata and Y. Ikada. 1988. "Oxidation of Polyethylene Surface by Corona Discharge and the Subsequent Graft Polymerization", *J. Polym. Sci., Part A, Polym. Chem.*, 26:3309−3322.

39. Noh, I., C.-H. Yoon and I.-J. Hwang. 1980. "Studies on Graft Copolymerization onto Polypropylene by Corona Discharge," *Inha University Report* (written in Korean), 7:99−107.

40. Oster, G. 1956. "Crosslinking of Polyethylene with Selective Wave Lengths of Ultraviolet Light," *J. Polym. Sci.*, 22; Oster, G. and Y. Mizutani. 1956. "Copolymers of Allyl Alcohol and Acrylonitrile Produced by Dye-Sensitized Photopolymerization," *J. Polym. Sci.*, 22:173−178.

41. Oster, G. and O. Shibata. 1957. "Graft Copolymer of Polyacrylamide and Natural Rubber Produced by Means of Ultraviolet Light," *J. Polym. Sci.*, 26:233−234.

42. Oster, G., G. K. Oster and H. Moroson. 1959. "Ultraviolet Induced Crosslinking and Grafting of Solid High Polymers," *J. Polym. Sci.*, 34:671−681.

43. Charlesby, A., C. S. Grace and F. B. Pilkington. 1962. "Crosslinking of Polyethylene and Paraffins by Ultra-Violet Radiation in the Presence of Sensitizers," *Proc. Roy. Soc.*, A268:205−221.

44. Takakura, K., G. Takayama and J. Ukida. 1965. "Ultraviolet-Induced Crosslinking of Poly(vinyl alcohol) in the Presence of Sensitizers," *J. Appl. Polym. Sci.*, 9:3217−3224.

45. Bellobono, I. R., F. Tolusso, E. Selli, S. Calgari and A. Berlin. 1981. "Photochemical Grafting of Acrylated Azo Dyes onto Polymeric Surfaces. 1. Grafting of 4-(N-ethyl, N-2-acryloxyethyl) Amino, 4'-Netro, Azobenzene onto Polyamide and Polypropylene Fibers," *J. Appl. Polym. Sci.*, 26:619−628.

46. Tazuke, S., T. Matoba, H. Kimura and T. Okada. 1980. In *Modification of Polymers*, C. E. Carraher and M. Thuda, eds., ACS Symposium Ser., No. 121, pp. 217−241.

47. Tazuke, S. and H. Kimura. 1978. "Surface Photografting. 1. Graft Polymerization of Hydrophilic Monomers onto Various Polymer Films," *J. Polym. Sci. Polym. Lett. Ed.*, 16:497−500.

48. Inoue, H. and S. Kohama. 1984. "Surface Photografting of Hydrophilic Vinyl Monomers onto Diethyldithiocarbamated Polydimethylsiloxane," *J. Appl. Polym. Sci.*, 29:877−889.

49. Ogiwara, Y., M. Kanda, M. Takumi and H. Kubota. 1981. "Photosensitized Grafting on Polyolefin Films in Vapor and Liquid Phases," *J. Polym. Sci., Polym. Lett. Ed.*, 19:457−462.

50. Ogiwara, Y., M. Takumi and H. Kubota. 1982. "Photoinduced Grafting of Acrylamide onto Polyethylene Film by Means of Two-Step Method," *J. Appl. Polym. Sci.*, 27:3743−3750.

51. Carlsson, D. J. and D. M. Wiles. 1976. "The Photooxidative Degradation of Polypropylene. Part I. Photooxidation and Photoinitiation Process," *J. Macromol. Sci. Rev. Macromol. Chem.*, C14(1):65−106.; Carlsson, D. J. and D. M. Wiles. 1974. "Photostabilization of Polypropylene III. Stabilizers and Macroketons," *Macromolecules*, 7:259−262.

52. Uyama, Y. and Y. Ikada. 1988. "Graft Polymerization of Acrylamide onto UV-Irradiated Films," *J. Appl. Polym. Sci.*, 36:1087−1096.

53. Somersall, A. C. and J. E. Guillet. 1975. "Photoluminescence of Synthetic Polymers," *J. Macromol. Sci., Revs. Macromol. Chem.*, C13:135−187.

54. Uchida, E., Y. Uyama and Y. Ikada. 1990. "A Novel Method for Graft Polymerization onto Poly(ethylene terephthalate) Film Surface by UV Irradiation without Degassing," *J. Appl. Polym. Sci.*, 41:677−687.

55. Ichijima, H., T. Okada, Y. Uyama and Y. Ikada. 1991. "Surface Modification of Poly(methyl methacrylate) by Graft Copolymerization," *Macromol. Chem.*, 192:1213−1221.

56. Kachan, A. A., L. L. Chervyastsova, K. A. Kornyev, E. F. Mertvichenko and N. P. Gnyp. 1967. "Investigation of the Kinetics and Mechanism of Radiochemical and Photochemical Grafting of Acrylonitrile from the Gaseous Phase to Polyamide Fibers," *J. Polym. Sci.*, C16:3033−3039.

57. Ogiwara, Y., M. Kanda, M. Takumi and H. Kubota. 1981. "Photosensitized Grafting on Polyolefin Films in Vapor and Liquid Phases," *J. Polym. Sci., Polym. Lett. Ed.*, 19:457−462.

58. Ogiwara, Y., K. Torikoshi and H. Kubota. 1982. "Vapor Phase Photografting of Acrylic Acid on Polymer Films: Effects of Solvent Mixed with Monomer," *J. Polym. Sci., Polym. Lett. Ed.*, 20:17−21.

59. Allmer, K., A. Hult and B. Ranby. 1988. "Surface Modification of Polymers. I. Vapour Phase Photografting with Acrylic Acid," *J. Polym. Sci., Polym. Chem.*, 26:2099−2111.

60. Ranby, B., Z. M. Gao. A. Hult and P. Y. Zhang. 1986. *Amer. Chem. Soc. Polym. Prep.*, 27(2):38.

61. Yao, Z. P. and B. Ranby. 1990. "Surface Modification by Continuous Graft Copolymerization. I. Photoinitiated Graft Copolymerization onto Polyethylene," *J. Appl. Polym. Sci.*, 40:1647−1661.

62. Goldblatt, R. D., J. M. Park, R. C. White and L. J. Matienzo. 1987. "Photochemical Surface Modification of Poly(ethylene terephthalate)," *Ame. Chem. Soc. Polym. Prep.*, 28:60−61.

63. Goldblatt, R. D., J. M. Park. R. C. White and L. J. Matienzo. 1989. "Photochemical Surface Modification of Poly(ethylene terephthalate)," *J. Appl. Polym. Sci.*, 37:335−347.

64. Lazare, S., P. D. Hoh, J. M. Baker and R. Srinivasan. 1984. "Controlled Modification of Organic Polymer Surfaces by Continuous Wave Far-Ultraviolet (185 nm) and Pulsed-Laser (193 nm) Radiation: XPS Studies," *J. Ame. Chem. Soc.*, 106:4288–4290.

65. Srinivasan, R. and S. Lazare. 1985. "Modification of Polymer Surfaces by Far-Ultraviolet Radiation of Low and High (Laser) Intensities," *Polymer*, 26:1297–1300.

CHAPTER 8

Surface Structures and Properties of Grafted Polymers

Surface modification of polymers by grafting of water-soluble chains has been studied extensively, as demonstrated in Chapter 7. However, in spite of the wide variety of possible applications of such modifications, the surface grafting technology has been applied only in a few cases on an industrial scale, probably because the basic and applied studies are still in their infancy. Another reason may be that such a hydrophilic, grafted polymer surface is relatively expensive to produce and is used mostly in aqueous environments since the presence of water is essential for the surface to exhibit its unique properties. Application fields where these grafted materials may be used in aqueous environments include marine science, biotechnology, and biomedical engineering. As these fields have attracted attention only in recent years, it is not surprising that polymer modification by surface grafting technology has not yet become common in industry.

8.1 STRUCTURES

8.1.1 Graft Length and Density

Grafted surfaces are very difficult to characterize because of the extremely low density of grafted chains on the surface. For instance, graft concentration amounts to only 0.1 $\mu g/cm^2$, if grafted chains with a molecular weight of 1×10^5 are assumed to be fixed as a monolayer at a density of one chain per 1 nm^2 on a surface. Conventional analytical methods are not sensitive enough to allow us to determine separately the length and the number density of grafted chains fixed on a surface at such a low concentration, although there are several methods to evaluate the overall graft concentration, which is a product of the chain molecular weight and the number density of the graft chains. Assuming that the molecular weight of the grafted chains is equal to that of the homopolymer chains produced simultaneously with the grafted chains in the region close to the substrate

139

surface, one can calculate the graft density from the graft concentration and the average molecular weight of the homopolymer.

Figure 8.1 shows a representative result for the graft polymerization of dimethylaminoethyl methacrylate (DMAEMA) onto the surface of a PET film by UV irradiation. Apparently, the molecular weight of the grafted chains remains constant, regardless of UV irradiation time, when the monomer concentration is fixed, whereas the graft density increases with the UV irradiation time. On the other hand, both the molecular weight and the density of grafted chains increased with the monomer concentration when the UV irradiation time was kept constant. This finding is in agreement with the conventional mechanism of radical polymerization, which predicts that the graft chain length can be readily reduced by addition of chain transfer agents to the aqueous monomer concentration, if necessary.

8.1.2 Highly Grafted Surface

The characterization described above was performed under the assumption that the surface has monolayered graft chains, but this assumption is unlikely if graft polymerization takes place to a high extent. In such a case, the surface structure with high graft densities can be assessed by XPS and SEM, although in a qualitative manner.

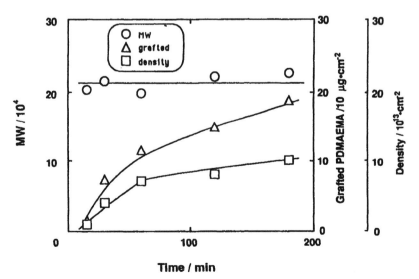

FIGURE 8.1. Variation of graft density of dimethylaminoethyl methacrylate onto PET film and molecular weight as a function of UV irradiation time.

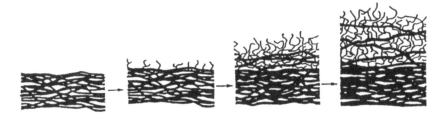

FIGURE 8.2. Schematic representation of physical structure of the grafted surface region.

A schematic presentation of the structure of grafted surface with different graft concentrations is given in Figure 8.2 [1]. The surface model with highly dense graft chains was depicted on the basis of XPS results of a PET film that underwent graft polymerization of AAm by UV irradiation. Even when the graft concentration was as high as 100 $\mu g/cm^2$, the XPS analysis showed that a small amount of PET component was always contaminated in the surface region. The illustration in Figure 8.2 may explain why such a high graft concentration is possible, although the graft polymerization must be localized on the surface. The significant resistance of graft chains to removal by mechanical rubbing of the graft surface, as well as the remarkable slipperiness of the grafted surface with a high graft concentration when hydrated, appears to support the surface structure illustrated in Figure 8.2.

8.1.3 SEM Observation of Lubricious Polymer Surface [2]

An SEM photograph of a grafted surface air-dried under ambient conditions did not show any significant difference from that of a nongrafted surface. However, the SEM photograph indicated a textured pattern when the grafted film was subjected to freeze-drying following hydration by immersion in water. Figure 8.3 shows a representative example of such textured surfaces for an ozonized polyurethane film grafted by immersion in an aqueous solution of *N,N*-dimethylacrylamide (DMAA) monomer in the presence of Mohr's salt after degassing. The textured structure was more clearly observed, as the graft density increased. It is interesting to note that the surface of the grafted films was transparent when air-dried, similar to that of virgin film, but turned opaque upon freeze-drying of the hydrated film. Interestingly, the dried opaque surface became transparent again without any change of textured structure when stored in air for a short time. The possible configurations of graft chains on a surface may be schematically represented as shown in Figure 8.4.

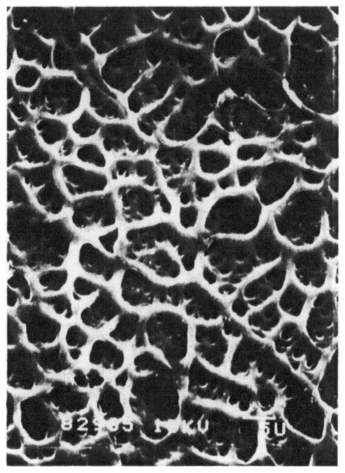

FIGURE 8.3. SEM photograph of the dimethylacrylamide-grafted polyurethane film. Film was freeze-dried after immersion in water. (Source: Inoue et al., 1992 [2].)

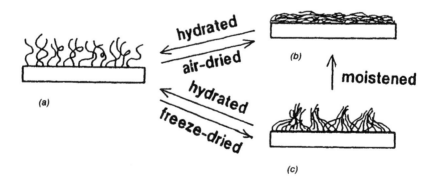

FIGURE 8.4. Schematic representation of the surface structure of polyurethane film grafted with water-soluble polymers in water (a), after air-drying (b) and after freeze-drying (c) [2].

The topography of the graft layer depends on the graft concentration as well as the surrounding environment, as demonstrated above. Ratner and his coworkers showed SEM photographs of a textured surface for low-density polyethylene films graft-copolymerized with 2-hydroxyethyl methacrylate and ethyl methacrylate by a radiation-induced technique [3,4]. However, the textured structure was quite different from that shown in Figure 8.3 in magnitude and appearance. Lazare et al. also showed a topography similar to Figure 8.3 for the PET film subjected to photoablation with a far-ultraviolet excimer laser [5,6]. When treated with ozone or exposed to UV, polyurethane exhibited no significant difference from the nontreated surface, indicating that significant erosion did take place in the course of the surface graft polymerization. It seems likely that the textured surface seen in Figure 8.3 is due to the buildup of closely lumped graft chains formed in the graft layer in the course of freezing, when the thickness of the graft layer and the graft density are relatively large. If such a grafted surface comes into contact with a large amount of water, the grafted layer is swollen, resulting in expansion of the grafted layer—which, however, cannot dissolve in the water as it is tethered to the water-insoluble substrate through covalent bonds.

The polyurethane surface grafted with water-soluble polymer chains can retain a high concentration of water molecules on the surface for a long period of time. Indeed, the freeze-dried surface was unstable in air even at room temperature, probably because of the immediate reorientation of the grafted polymer chains through quick absorption of water from the atmosphere. Although the grafted surface may interact less strongly with another solid surface than the ungrafted surface, it does not become slippery unless water is present. This lack of slipperiness is due to the fact that there is no lubricant at the interface between the two solid surfaces.

8.2 SURFACE PROPERTIES

Polymer surfaces with grafted polymer chains exhibit many unique properties that conventional polymer surfaces without graft chains do not have. Some examples of such properties will be described below.

8.2.1 Protein and Cell Interactions

The first event that takes place on a polymer surface when brought into contact with a biological system is generally protein adsorption, followed by cell attachment. Unless any protein is present, cells and microorganisms attach to the polymer surface. This protein adsorption and attachment of biological components trigger a subsequent series of biological reactions toward the polymeric materials. Therefore, the regulation of this first event is of prime importance to controlling the biological reactions. It has been found that the polymer surface that is grafted appropriately with nonionic, water-soluble polymer chains minimizes the protein adsorption and cell adhesion. Figure 8.5 shows a result for a polyurethane film grafted with PAAm [7]. It is seen that excessive grafting results in rather remarkable protein adsorption. It should be noted that such minimum protein adsorption has been well known for a long time for the hydrogels used for protein analysis, such as PAAm gel for electrophoresis, sephadex for protein gel filtration, and soft agar for cell culture. It is likely that the surface structure of these hydrogels resembles that of the grafted surface described above.

8.2.2 Protein Immobilization

The polymer carriers used most extensively for protein immobilization may be sepharose and agarose beads. As these hydrogels probably have tethered, water-soluble chains on the surface, it is reasonable to think that the grafted surface can provide a good substrate for protein immobilization if the graft chains have functional groups to be used for the protein coupling reaction. For this purpose, one can graft-polymerize monomers having acyl chloride, carboxyl, hydroxyl, amino, or epoxy groups like

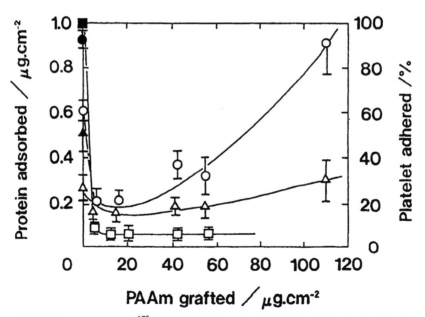

FIGURE 8.5. Adsorption of ^{125}I-labeled proteins (IgG, BSA) and adhesion of platelet to the polyurethane film grafted with polyacrylamide (PAAm). ○, ●: IgG; △, ▲: BSA; □, ■: platelet. Open marks—grafted with PAAm, closed marks—ungrafted [7].

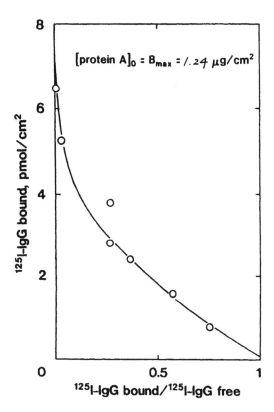

FIGURE 8.6. Scatchard plot of the binding of ^{125}I-human IgG to protein A-immobilized PET fibers. 1 ml of ^{125}I-human IgG in 0.1 M PBS was added to PET fibers immobilized with 0.16 g/cm^2 of protein A.

acryloyl chloride, acrylic acid, methacrylic acid, HEMA, AAm (subjected to Hoffman degradation, Mikel's reaction, or alkaline hydrolysis), and glycidyl methacrylate onto a polymer surface. We have already covalently immobilized different proteins onto a polymer surface grafted with polyacrylic acid—collagen for cell-adhesive surfaces, enzymes for biosensors, and immunoglobulin for immunoadsorbents. Figure 8.6 shows the result of immunogammaglobulin G sorption to a PET fabric immobilized with protein A through surface graft polymerization of acrylic acid [8].

8.2.3 Adhesion of Grafted Surfaces

Recently we have found that a polymer surface modified by graft polymerization of water-soluble monomers undergoes substantial adhesion to another surface when they are brought into contact in the presence of water under pressure and subsequently dried. A PET film was used for this study after surface grafting with water-soluble monomers by a photoirradiation technique in the presence of riboflavin after plasma treatment. The adhesive force between the two polymer films was measured after wetting them with water, lapping them together under a load, and then allowing them to spontaneously dry [9].

Figure 8.7 shows logarithmic plots of shear strength between the two identical AAm-grafted films with different graft densities, against the drying time. It appears that the graft density does not have a large effect on the adhesive strength, as long as it is larger than 10 $\mu g/cm^2$. This effect of graft density suggests that the adhesive force is governed by the water content in the grafted layer and the physical entanglement of water-soluble grafted chains. In other words, grafted chains will function as adhesives when the surface-grafted film is brought into contact with another film in the presence of water and the water is removed from the interface by simple evaporation. If the opposing surface also possesses grafted chains on the surface, some chain entanglement between the polymer molecules will take place, resulting in the stronger adhesion, as depicted in Figure 8.8.

8.2.4 Lubrication of Grafted Surfaces [10]

The adhesion study described above suggests that the surface with water-soluble grafted chains can absorb abundant water, which may enhance the

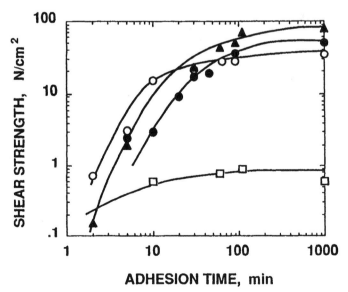

FIGURE 8.7. Effect of the graft density on the shear strength between two identical AAm-grafted PET films. Graft density ($\mu g\ cm^{-2}$): □—1; ○—10; ●—50; ▲—90 [9].

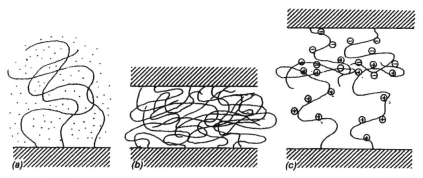

FIGURE 8.8. Schematic representation of water-soluble graft chains on the film surface. (a): Diffused graft chain before drying; (b): mechanically entangled graft chains; (c): polyion-complexed graft chains with opposite charges.

lubricity between the substrate and the opposing surface, functioning as fluid film. To confirm this assumption, a study was performed focusing on the role of grafted chains by measuring the friction between a surface-grafted film and glass or stainless steel in a wet state. Polyetherurethane was employed as the substrate polymer and N,N-dimethylacrylamide (DMAA) was used as the nonionic water-soluble monomer. The ozone method was chosen for graft polymerization of monomer onto the surface of the substrate polymer because of its relatively simple procedure.

The coefficient of friction (μ) of films was determined against a cleaned steel in distilled water, both for the static friction (μ_s) and kinetic friction (μ_k). The μ_s value was determined in two different manners: either by measuring a sliding angle or a static frictional force using a modified ASTM apparatus as illustrated in Figure 8.9. A columnar steel having a radius of 10.0 mm and a mass of 9.0 g was placed on a test film and allowed to travel over the film at a sliding speed of 0.2 mm/s. The force needed just for the columnar slider to start moving was calculated from the displacement of a leaf spring with the aid of a laser beam detector. As it was difficult to determine the μ_k values with this apparatus over a wide range of sliding speeds, a thrust-collar type apparatus (shown in Figure 8.10) was utilized for the dynamic friction measurement. A steel cylinder with a ring-shaped test film attached was allowed to rotate against a glass plate at different velocities of rotation. During rotation, the cylinder was supported with a sphere made of silicon nitrite to keep it in the horizontal position and provide homogeneous contact of the specimen with the glass plate. The frictional force was determined by measuring the tension applied to the thread connected to the leaf spring.

When the ring-shaped specimen (inner diameter r_i and outer diameter r_o) is rotated under the load P, the mean pressure P_m on the specimen is given by Equation (8.1):

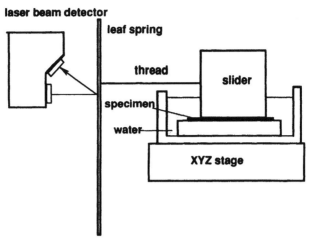

FIGURE 8.9. Schematic representation of the apparatus for determining a coefficient of static friction by the modified ASTM method. Slider: diameter = 1.0 cm, mass = 9 g.

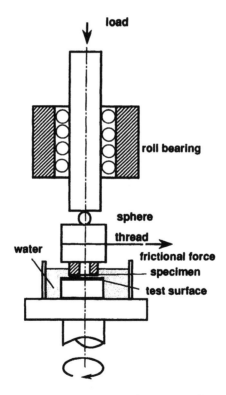

FIGURE 8.10. Thrust-collar type apparatus for the kinetic friction measurement.

$$P_m = P/\pi \, (r_o^2 - r_i^2) \qquad (8.1)$$

The mean radius r_m of the specimen is

$$\pi r_m (r_o^2 - r_i^2) = \int_I^O r \, 2\pi r dr \qquad (8.2)$$

when the contribution of area is considered. The coefficient of friction can be calculated by Equation (8.3):

$$\mu = M/Pr_m \qquad (8.3)$$

where M corresponds to the friction moment. The outer and inner diameters of the ring-shaped specimens used in this study were 0.45 and 0.20 mm, respectively.

The μ_s of surface-grafted polyurethane films varied depending on the presoaking time in distilled water. The grafted polyurethane film having a graft density of 500 $\mu g/cm^2$ was completely dried and then immersed in distilled water at 25°C. The μ_s value calculated with the sliding angle method was plotted against the immersion time in Figure 8.11. As can be clearly seen, μ_s decreased continuously with the immersion time up to about 1 h. This indicates that it takes a considerably long period of time for the grafted film to get a fully hydrated surface.

The μ_s value of the grafted film depended also on the contacting time of the surface placed on the counterpart under a load. A columnar steel was carefully placed on a surface-grafted, fully hydrated polyurethane film without applying an additional load on the grafted film, and μ_s was measured after a determined period of time with the apparatus shown in Figure 8.9. The results are shown in Figure 8.12 for surface-grafted films having

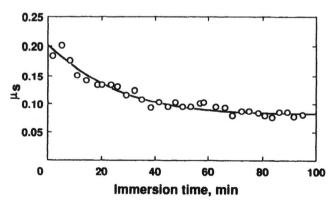

FIGURE 8.11. Variation of the coefficient of static friction (μ_s) with the immersion time in water for a surface-grafted polyurethane film.

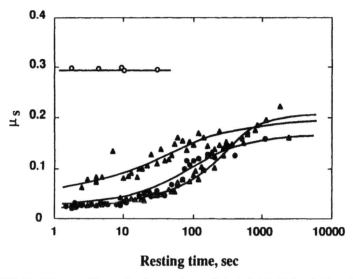

FIGURE 8.12. Influence of the resting time on the coefficient of static friction (μ_s) in water for surface-grafted polyurethane films. Graft density (μg/cm^2): ○—0, ●—33, △—100; ▲—340. Contact pressure: 1.0 kPa.

various graft densities. It is clearly seen that for all the polyurethane films, μ_s increased with the resting time, although the μ_s of ungrafted film exhibited a constant value around 0.3, irrespective of resting time. The increase in frictional force with time was more clearly observed when an additional load was added on the grafted surface during frictional traveling. Figure 8.13 shows the transition of frictional force with sudden change of load from 1.2 to 2.2 N on the polyurethane film having a graft density of 350 μg/cm^2. Frictional force increased gradually with time for a few minutes until it reached a new equilibrium state.

The variation of μ_k is plotted in Figure 8.14 against the sliding speed for fully hydrated grafted films of two different graft densities. As can be seen, the μ_k value of two films varied with the sliding speed in different manners; μ decreased monotonously for the ungrafted or the grafted film with lower graft density, whereas the polyurethane film with the graft density of 350 μg/cm^2 exhibited a maximum μ.

In order to study the adhesion between a grafted film and a glass surface in water, the adhesion force was measured using the apparatus illustrated in Figure 8.15. A grafted film was attached to the bottom of a water bath placed on a stage, and a glass ball was suspended from a leaf spring. The adhesive force between the film and the glass ball in water was measured at 25°C by the use of a laser beam detector. It was found that the adhesive force increased with the contacting time (Figure 8.16) as well as with the contact-

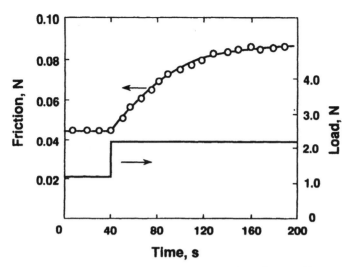

FIGURE 8.13. Transition of friction with sudden change of load from 1.2 N to 2.2 N. Graft density: 350 $\mu g/cm^2$; sliding speed: 2.1 mm/s.

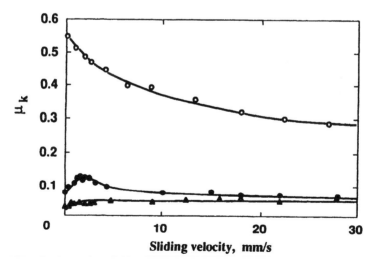

FIGURE 8.14. Variation of the coefficient of kinetic friction (μ_k) with the sliding velocity for surface-grafted polyurethane films in water. Graft density ($\mu g/cm^2$): ○—0, ●—140, △—350.

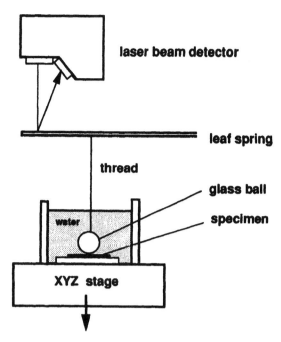

FIGURE 8.15. Schematic representation of the apparatus for measurement of the adhesive force between a test specimen and a glass ball in water.

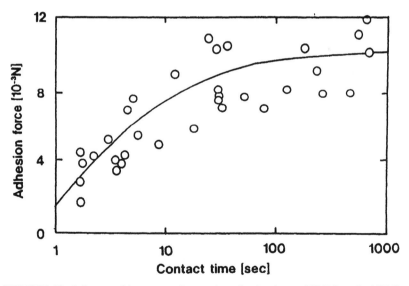

FIGURE 8.16. Influence of the contact time on the adhesion force of DMAA-grafted PU film against glass ball.

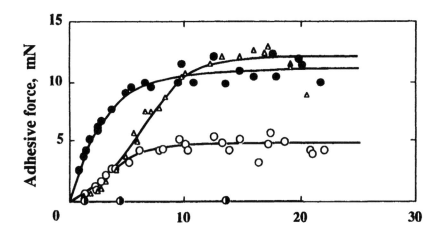

Contacting force, mN

FIGURE 8.17. Relation between the contacting force and the cohesive force for surface-grafted polyurethane films. Graft density (μg/cm^2): ◑—0; ○—33; ●—140; △—340.

ing force and the graft density, so long as the contacting force was larger than 10 mN (Figure 8.17).

The results described above indicate that, upon contact with water (as shown in Figure 8.11) the grafted surface does not instantaneously become fully hydrated and, even after full hydration, gradually loses water molecules under a load at the contacting region, leading to an increase both in the frictional force and in the adhesive force. These findings seem to be analogous to the squeeze membrane effect in lubrication of quasi-hydrodynamic flow, but, as can be seen from Figure 8.12, the time scale is small compared with that of ordinary hydrodynamic lubrication, probably because of the restricted motion of water molecules solvated to the water-soluble graft chains. As the graft density increases, lots of water molecules will be absorbed which play an important role as fluid-film lubricant. As the Stribeck plot (Figure 1.8 in Chapter 1) indicates, the variation of μ_k with the increasing velocity or viscosity may probably be caused by the mixed boundary and fluid-film lubrication. It is likely that the fluid-film lubrication becomes more predominant for the film with higher graft density.

8.2.5 Drag Reduction

It would be interesting if the lubricating surface described so far could be applied, for instance, to the wall of a pipe transporting fluid or to the surface of a boat or an aircraft to increase the volume flow of fluid or the craft speed. In 1948, Toms [11] reported the flow of linear polymer solutions through tubes at large Reynolds numbers. In his experiment, the frictional drag in the turbulent flow of chlorobenzene decreased when a small amount of PMMA was added in the solvent. Today, the phenomenon of such a "drag reduction" system is often called the "Toms effect." Application of this phenomenon was first directed to the oil transportation pipeline industry as it had the potential for reducing the energy required for oil delivery. In this case, injection of only a small concentration of high-molecular-weight polymers in the turbulent flow range led to a substantial drag reduction in the pipelines. Interthal and Wilski [12] demonstrated a pressure drop along a pipeline of 3000 m length filled with flowing water by adding polyacrylamide, and showed that an addition of 30 ppm polyacrylamide could reduce the total pressure loss by almost half compared to the flow without PAAm. However, for longer pipelines it was no more applicable because the added polymer underwent significant degradation due to the thermal and mechanical deterioration over the long distance. This problem would be easily overcome, provided additives consisting of some nonorganic materials are found which would still show a significant effect on drag reduction. Various flow-improving additives used to date are listed in Table 8.1.

The drag reduction has been accounted for in terms of decreasing turbulence, the wall effect (not related to the surface nature of the pipe, but to the shear thinning wall layer), the adsorption effect, stretching of polymer molecules in flow, and so forth. A detailed description of the drag reduction phenomenon can be found in a review article written by Andreis et al. [13].

The surfaces of many fishes, such as eels, are very slippery in the presence of water, and marine mammals like porpoises move through the water at high speed. Scientists have tried to learn how these animals can attain such a slippery surface and how they can swim through water avoiding frictional drag. As a result, several techniques for drag reduction systems have been proposed which imitate the skins of natural animals, since such systems have been highly prized in the fields of marine science, aerodynamics, fluid transportation, and biomedical engineering. It has often been postulated that a ship with a surface identical to that of a dolphin's skin could attain a similar sailing speed. Kraemer [14] had tried to apply an elastic material to a ship-like model and evaluate the effect, measuring the drag coefficient between the ship surface and seawater. The size and structure he employed are shown in Figure 8.18. Although he concluded that a drag reduction effect was observed in the restricted range of Reynolds number (Figure 8.19), the effect could not be found over the wide range of the turbulent flow.

On the other hand, many experimental results indicate that riblets are effective in reducing the skin friction drag [15]. Walsh intensively studied

TABLE 8.1. Flow Improving Additives Used to Date [10].

Solid Medium	Fibrous Medium
Fine-grained sands	Wood
Small glass beads	Nylon
Flax grains	Peat
Coal dust	Rayon
Lubricating grease, corundum	Asbestos, glass

Micelle-Forming System	
Tri-*n*-butyl-tin(IV) fluoride, cetyltrimethylammoniumbromide, cationic, anionic, and non-anionic surfactants	

Synthetic Organic Polymers	
Polymethylmethacrylate (PMMA)	Polyethyleneoxide (PEO)
Polyisodecylmethacrylate	Polystyrene
Polyacrylic acid (PAA)	Polystyrenesulfonate
Polyacrylamide (PAAm)	Polyethylenimine (PEI)
Hydrolyzed polyacrylamide	Polyvinylalcohol (PVA)
Glyoxylyzed polyacrylamide	Polyvinylpyrrolidone (PVP)
Polyisobutylene (PIB)	Poly-*cis*-isoprene (PCIP)

Synthetic, Inorganic Polymers	
Polydimethylsiloxane (PDMS)	Polyphosphate

Biological Additives	
Guar gum (GG)	Guarantriacetat (GTA)
Carboxymethylcellulose (CMC)	Hydroxyethylcellulose (HEC)
Xanthan	Fish mucus
Chlorella stigmataphara	*Porphyridium aerugineum*
Porphyridium cruentum	*Chaetoceros didynuis*
Protocentrum micans	*Arthrobacter viscosus*
Chaetoceros affinis	*Exuviella casszbuca*
Anabaena flos-aquae	*Chlamydenas peterfii*
Pseudomonas and *Neisseria* species	

the drag reduction of turbulent boundary layer using riblets [16–18] and concluded that the maximum drag reduction by the V-groove riblet was dependent on the height and spacing of the riblets in law-of-the-wall variables, regardless of the free-stream Reynolds number or upstream boundary-layer history. The V-groove riblet having the optimal structure in height and spacing exhibited 7 to 8% drag reduction (Figure 8.20). Riblet materials [19] and a ciliated material [20] were also disclosed in patents. Gould and Kliment [21] reported that water flow through a tube treated with

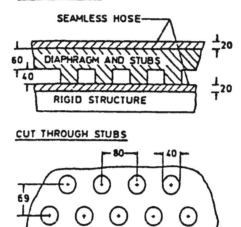

FIGURE 8.18. The dimensions of the ducted rubber coating tested (dimensions in 1/1000 of an inch) [14].

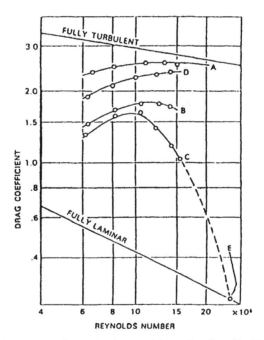

FIGURE 8.19. The drag coefficient of various models as a function of the Reynolds number. Curve A—the rigid reference model. Curves B, C, and D—fully coated models with a stiffness of the coating on the cylindrical section: B—1600 PCI, C—800 PCI, D—600 PCI. Point E—Reynolds number for best performance at 800 PCI stiffness: calculated from simplified theory for ducted coatings [14].

156

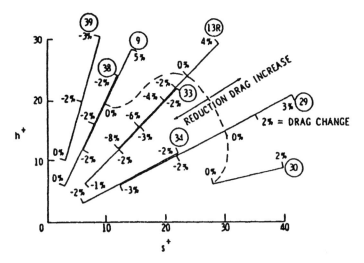

FIGURE 8.20. Drag reduction region of h^+, s^+ for symmetric V-groove riblets. (Copyright ©AIAA 1982. Used with permission.)

hydrophilic polyurethane for surface lubricity was found to increase the flow by 22% as compared to the flow through the untreated tube.

We also studied the effect of surface lubricity on the drag reduction using a surface-grafted HDPE. In order to maintain a constant turbulent flow over a wide range of Reynolds numbers for a long period of time, a large-scale flowing system was constructed as shown in Figure 8.21. The grafted HDPE film was attached to the inner wall of the test pipe (I.D. = 19.2 cm,

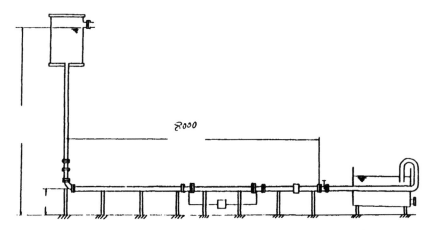

FIGURE 8.21. Flowing system to determine a friction factor of test pipe.

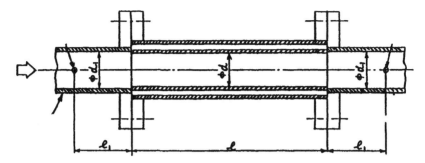

FIGURE 8.22. Test piece for determining the friction factor in turbulent flow.

length = 60 cm) inserted in the midst of the flow path, as depicted in Figure 8.22. Flow test was conducted with the turbulent flow at the Reynolds number $(1.0 - 7.0) \times 10^4$ and a flow rate of $0.7 - 4.5$ m/s. Friction factor λ was calculated based on the Prandtl-Karmer equation:

$$1/\sqrt{\lambda} = 2.0 \log Re\sqrt{\lambda} - 0.8 \tag{8.4}$$

The walls of the smooth pipe with and without the grafted film both obeyed exactly the Prandtl-Karmer equation, as can be seen in Figure 8.23. It indicates that the attached film influences neither interference nor drag reduction of the liquid flow. It is likely that the surface structure of the wall does not play a significant role in drag reduction, because the slip between the wall surface and the fluid is insignificant. The coefficient of friction of

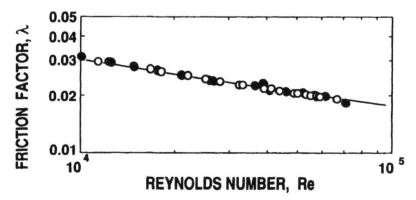

FIGURE 8.23. Plots of friction factors calculated for the inner wall of a pipe untreated ($\bigcirc$) and surface-grafted with acrylamide ($\bullet$) at various Reynolds numbers. Straight line shows the Prandtl-Karmer equation.

the surface-grafted film did not change even after the flow test was run for more than 60 hours at a flow rate of 3.4 cm/s, Reynolds number of 6.4 × 10^4, and shear wall stress of 30 N/m^2.

On the other hand, a polyurethane tube (I.D. = 0.65 mm length = 200 cm) whose inner wall was surface-grafted with acrylamide and DMAA, always retarded the laminar flow (Reynolds number less than 100). It is known that the surfaces of boats and bathing suits, treated with a mucous material, can achieve a higher speed at the same energy. This increase in speed is not due to the wall slip, but to removal of the mucous material in the course of movement. Although a similar mucous layer exists at the region between the grafted surface and water, the grafted chains are tethered to the solid substrate through a covalent bond inhibiting removal of the mucous layer.

Since the pressure drop associated with flow in a pipe or tube may be caused by many factors such as shape (especially at the inlet part), curvature, temperature, and contaminants, absolutely correct data are difficult to obtain by such a simple experiment. However, simply assuming that the flow retardation is caused by narrowing of the inner diameter of the tube due to the swelling of the graft layer, one can use Hagen-Poiseuille's equation to check the validity of this correlation under laminar flow. The effective diameter calculated based on the narrowing of the tube's inner diameter – due to the swelling of the graft layer – was several micrometers. This value was in agreement with that estimated from the thickness of the swollen graft layer obtained from the staining.

8.2.6 Antistatic

The generation of electrostatic charges on polymers has been a nuisance in many fields associated with polymer and textile technologies. In addition to the discomfort arising from static cling or static discharge, more serious problems include damage to semiconductors and sometimes to human life through fire or explosion. Therefore, it would be of great value to devise a simple method for eliminating static generation during the manufacturing process or to develop a new polymer that prevents charge generation. The mechanism by which the triboelectric charge is generated is still controversial and it remains unclear whether the electric charge carriers are electrons, ions, or both. There are many complicating factors that influence the absolute value of a triboelectric charge generated on a polymer surface. For instance, atmospheric conditions such as temperature and humidity will play a great role in the electrification of a surface.

Most industrial films, plastics, and rubbers are hydrophobic and are readily charged by rubbing. Since less hydrophobic materials are generally

TABLE 8.2. Electrostatic Properties of UV-Irradiated and Surface-Grafted Polymers.

Specimen	UV Irradiation Time (h)	Triboelectric[a] Charge (kV)	Half Decay[b] Time (s)		% Decay[b] after 3 min.	
			+5 kV	−5 kV	+5 kV	−5 kV
PS	0	−5.5	VL	VL	0.1	0.1
PS	1	−2.3	VL	VL	0.2	0.2
PS	2	−1.0	VL	VL	0.2	1.2
PS	5	−0.9	VL	VL	2.7	2.5
PET	0	−3.7	VL	VL	0.1	0.3
PET	1	−3.3	VL	VL	0.2	0.3
PET	2	−3.2	VL	VL	0.2	0.2
PET	5	−1.0	VL	VL	0.2	0.2
PET-g-DMAA	—	−0.5	18	21	78.0	80.0
PS-g-AA	—	−0.4	255	249	31.0	38.2

DMMA, N,N-dimethyl acrylamide; AA, acrylic acid.
[a] 25°C, 60% RH.
[b] 20°C, 75% RH. VL, very long (longer than 1 hour).

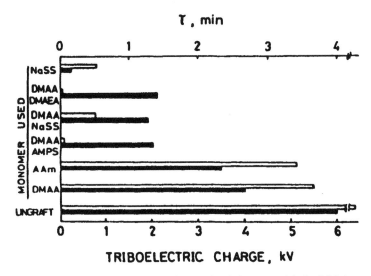

FIGURE 8.24. Electrostatic properties of the poly(ethylene terephthalate) fabrics surface-grafted with various monomers. Triboelectric charge generated upon rubbing with a cotton fabric is given in absolute values neglecting the sign of the charge. Half decay time (τ) is the average of values observed when +5 kV and −5 kV were applied. NaSS, sodium styrenesulfonate; DMAEA, *N,N*-dimethylaminoethyl acrylate; AMPS, 2-acrylamido-2-methylpropane sulfonic acid; ▭, half decay time; ▬, triboelectric charge [22].

more difficult to charge, it is interesting to study triboelectric charging of hydrophobic polymeric materials whose surfaces are grafted with hydrophilic polymer chains. A PET film was graft-polymerized with DMAA by the photoirradiation technique, as described in Section 7.2.6.3. The triboelectric charge of surface-grafted PET film is given in Table 8.2 [22]. The charge generated upon rubbing at 25°C and 60% RH was −0.5 kV, and the half decay times when +5 kV and −5 kV are initially applied are as short as 18 and 21 seconds, respectively. In contrast to the very slow decay for the other UV-irradiated samples in Table 8.2, the decay time was found to decrease considerably with an increase in the amount of PAAm grafted. It is expected that if the graft chain possesses cationic or anionic groups, the surface-grafted material should become more effective in terms of its antistatic property, as is shown for the triboelectrification and decay of charges in Figure 8.24 for PET fabrics that were surface-grafted with various water-soluble monomers. The half decay time of the ungrafted PET fabrics was around 4 hours. One can clearly conclude, based on the data in Figure 8.24 compared to data measuring just a few minutes when the PET was grafted with AMPS, that electrostatic charging is drastically inhibited by graft polymerization of ionic monomers.

8.3 REFERENCES

1. Uchida, E., Y. Uyama and Y. Ikada. 1990. "XPS Analysis of the Poly(ethylene terephthalate) Film Grafted with Acrylamide," *J. Polym. Sci., Part A, Polym. Chem.*, 28:2837–2844.

2. Inoue, H., Y. Uyama, E. Uchida and Y. Ikada. 1992. "Scanning Electron Microscope Observation of Lubricious Polymer Surface for Medical Use," *Cells and Materials*, 2:21–28.

3. Ratner, B. D. 1985. In *Surface and Interfacial Aspects of Biomedical Polymers*, J. D. Andrade, ed., New York: Plenum, pp. 373–394.

4. Ko, Y. C., B. D. Ratner and A. S. Hoffman. 1981. "Characterization of Hydrophilic–Hydrophobic Polymeric Surfaces by Contact Angle Measurements," *J. Coll. Interface Sci.*, 82:25.

5. Lazare, S. and R. Srinivasan. 1986. "Surface Properties of Poly(ethylene terephthalate) Film Modified by Far-Ultraviolet Radiation at 193 (laser) and 185 nm (low intensity)," *J. Phys. Chem.*, 90:2124–2131.

6. Novis, Y, J. J. Pireaux, A. Brezini, E. Petit, R. Caudano, P. Lutgen, G. Feyder and S. Lazare. 1988. "Structural Origin of Surface Morphological Modifications Developed on Poly(ethylene terephthalate) by Excimer Laser Photoablation," *J. Appl. Phys.*, 64:365–370.

7. Fujimoto, K., H. Tadokoro, Y. Ueda and Y. Ikada. 1993. "Polyurethane Surface Modification by Graft Polymerization of Acrylamide for Reduced Protein Adsorption and Platelet Adhesion," *Biomaterials*, 14(6):442–448.

8. Kato, K. and Y. Ikada. 1991. "Immobilization of Protein A onto the Surface of Microfibers," *The 13th Ann. Meeting Japanese Soc. Biomater. Prep., Kyoto*, p. 69.

9. Chen, K.-S., Y. Uyama and Y. Ikada. 1992. "Adhesive-Free Adhesion of Grafted Surfaces with Different Wettabilities," *J. Adhesion Sci. Technol.*, 6(9):1023–1035.

10. Ikeuchi, K., T. Takii, H. Norikane, N. Tomita, T. Ohsumi, Y. Uyama and Y. Ikada. 1993. "Water Lubrication of Polyurethane Grafted with Dimethylacrylamide for Medical Use," *Wear*, 161, 179–185; Uyama, Y., T. Takii, H. Norikane, R. Sekine, K. Ikeuchi and Y. Ikada. 1992. "Tribology of Surface-Grafted Polymers for Medical Use," *7th Intn'l Conf. Biomedical Eng., Singapore, Dec. 1992*.

11. Toms, B. A. 1948. *Proc. 1st Intern. Congr. on Rheology*.

12. Interthal, W. and H. Wilski. 1985. "Drag Reduction Experiments with Very Large Pipes," *Colloid Polym. Sci.*, 263:217–229.

13. Andreis, M., H. Grager, J. L. Koenig, M. Kotter and W.-M. Kulicke. 1989. *Advances in Polym. Sci.*, 89:1–68.

14. Kraemer, M. O. 1960. *J. Ame. Soc. Nav. Eng.*, 72.

15. Choi, K.-S., 1987. "On Physical Mechanisms of Turbulent Drag Reduction Using Riblets," *2nd Int. Symp. Transport Phenomena in Turbulent Flows, Tokyo, Oct. 25–29*, pp. 173–186.

16. Walsh, M. J. 1982. "Turbulent Boundary Layer Drag Reduction Using Riblets," *AIAA Report No. 82-0169*, pp. 1–8.

17. Walsh, M. J. 1983. "Riblets as a Viscous Drag Reduction Technique," *AIAA J.*, 21(4):485–486.

18. Walsh, M. J. and A. M. Lindemann. 1984. "Optimization and Application of Riblets for Turbulent Drag Reduction," *AIAA Report No. 84-0347*, pp. 1–10.

19. Marentic, et al. Japan patent 61-278500, Dec. 9, 1986.
20. Technology Resources Inc. Japan patent 61-175200, Aug. 6, 1986.
21. Gould, F. E. and C. K. Kliment. U.S. patent 4810543, 1989.
22. Uyama, Y. and Y. Ikada. 1990. "Electrostatic Properties of UV-Irradiated and Surface Grafted Polymers," *J. Appl. Polym. Sci.*, 41:619–629.

ABBREVIATIONS

Polymers

EVA	ethylene-vinyl acetate copolymer
EVAL	ethylene-vinyl alcohol copolymer (VAECO)
HDPE	high-density polyethylene
LDPE	low-density polyethylene
PAAm	poly(acrylamide)
PDMS	poly(dimethyl siloxane)
PE	polyethylene
PEO	poly(ethylene oxide)
PEG	poly(ethylene glycol)
PET	poly(ethylene terephthalate)
PHEMA	poly(2-hydroxyethyl methacrylate)
PMMA	poly(methyl methacrylate)
PP	poly(propylene)
PTFE	poly(tetrafluoroethylene)
PVA	poly(vinyl alcohol)
PVC	poly(vinyl chloride)
PVdC	poly(vinylidene chloride)
PVP	poly(N-vinyl pyrrolidone)
VAECO	ethylene vinyl-alcohol copolymer (EVAL)

Monomers

AAm	acrylamide
DMAA	N,N-dimethylacrylamide
DMAEMA	dimethylaminoethyl methacrylate
DMAP	dimethyl aminopyridine
EHA	2-ethylhexyl acrylate
EMA	ethyl methacrylate

HEMA   2-hydroxyethyl methacrylate
NVP   *N*-vinyl pyrrolidone
MAA   methacrylic acid

Others

AIVC   4,4′-azobis-4-cyanovaloil chloride
ATR   attenuated total reflection
BP   benzophenone
BSA   bovine serum albumin
CVD   chemical vapor deposition
DCC   dicyclohexylcarbodiimide
DMF   dimethyl formamide
FWHM   full width at half maximum
HMDI   hexamethylene diisocyanate
IOL   intraocular lens
IRE   internal reflection element
MEK   methyl ethyl ketone
PTF   precorneal tear film
SEM   scanning electron microscope
SIMS   secondary ion mass spectroscopy
XPS   X-ray photoelectron spectroscopy

abrasion, 14
acetylene, 79, 81
acrylamide, 52, 60, 67, 92, 113, 116,
        117, 118, 120, 123, 124,
        125, 126, 128, 131, 158
acrylic acid, 52, 53, 61, 69, 114, 115,
        116, 117, 121, 123, 125,
        126, 131, 145
acrylonitrile, 61, 65, 82, 115, 119, 130
active hydrogen, 49, 50
adhesion, 8, 14, 36, 144, 145, 150
advancing contact angle, 98, 101
agarose, 51, 144
albumin, 51
aneurysm, 56
anti-apolipoprotein, 88
antibody, 87, 88
antifogging, 75, 76
antifouling, 46, 75, 76
antimicrobial agent, 58
antithrombogenicity, 76
arteria femoralis, 57
artificial joint, 13, 28, 32, 42, 61
ASTM, 17, 22, 23, 26, 48, 147
ATR-IR, 78, 84, 92, 102, 107

beeswax, 66
benzophenone, 124, 125, 126, 130, 131
benzoyl peroxide, 114, 130
bioactive compound, 88
biocompatibility, 55, 76
biodegradable/bioabsorbable, 64, 65,
        67
biosensor, 145
blood plasma, 23
body cavity, 58, 118
bone cement, 62
Bowden-Leben friction test, 22
British Standard (BS), 17

Brownian motion, 73
buoyancy factor, 98

catgut, 63
catheter, 26, 28, 52, 55, 56, 57, 58, 59,
        60, 61, 67, 69
cathetometer, 36
caustic soda, 78
cellulose, 43, 45, 50, 74, 78, 79, 88,
        93, 95, 113, 124
    methyl, 34, 65
    carboxy methyl, 34, 50
    hydroxy propyl, 34, 50, 69
ceric ammonium nitrate, 113
ceric ion, 113
chain scission, 85
chemical shift, 93, 104
chondroitin sulfate, 51
chromic acid, 82, 93
chromophore, 127, 128
ciliated material, 155
Co^{60}, 117, 118
coefficient of friction, 2, 3, 5, 17, 22,
        23, 26, 27, 30, 32, 33, 34,
        42, 43, 44, 47, 48, 49, 67,
        69, 77, 147, 158
collagen, 28, 63, 64, 113, 145
compressive yield value, 5
contact angle, 33, 43, 77, 78, 82, 84,
        91, 95, 96, 97, 98, 111, 115,
        128
contact angle hysteresis, 97, 100
corona discharge, 53, 79, 83, 84, 85,
        91, 123, 128
Coulomb, 2
creep, 13
cross link, cross linking, 41, 42, 47,
        52, 53, 65, 79, 81, 116, 123,
        125

cure, curing, 47, 48, 53
cyanogen bromide, 113

delamination, 50
depth profile, 73, 104
derivatization, 93, 94
dextran, 112, 113
DMAA, 59, 141, 147, 152, 159, 160
drag reduction, 46, 153, 154, 155, 157, 158

E.P. additives, 10, 13
electric discharge, 82
electron beam, 53, 79, 115
elongation at break, 14
endothelium, 35, 36, 57
enzyme, 145
epoxy resin, 83
ethylene-vinyl alcohol copolymer
        (EVAL), 112
ethylene-vinyl acetate copolymer
        (EVA), 52, 59, 60, 118, 126, 127

far UV, 131, 132
free energy, 91, 118
free radical, 81, 118, 119
freeze-drying, 141, 142, 143
friction factor, 157, 158
friction moment, 149
FWHM, 93

glow discharge, 23, 78, 79, 81, 82, 83,
        91, 118, 119, 121, 128
glycidyl methacrylate, 145

Hagen-Poiseuille's equation, 159
half decay time, 160, 161
hard segment, 47
HEMA, 68, 69, 92, 116, 117, 118,
        125, 143, 145
high-density polyethylene (HDPE), 32,
        41, 85, 101, 131, 157
hydrocolloidal dispersion, 51
hydrogen abstraction, 124
hydroperoxide, 123, 126, 128

immune complex, 88
immunoadsorbent, 88
immunogammaglobulin, 145
interpenetrating polymer network, 46, 50

interpolymer, 48, 50, 51
intraocular lens (IOL), 35, 65
introducer sheath tube, 60
iodometry, 123, 132
ion beam, 79, 105
ionene (ionic amine), 51
isocyanate, 47, 48, 49, 50, 51, 58, 78, 113

kink, 57
knot, 37, 63, 64, 65

latex, 47, 61
law-of-the-wall variable, 155
LDPE, 78, 84, 93, 125, 131, 141
lecithin, 64
Lidocaine jelly, 55
lipid, 64
LoFric®, 59
lubrication
        boundary lubrication, 9, 10, 33
        fluid-film lubrication, 8, 11, 33,
                45, 61
        solid-film lubrication, 1, 8, 13, 67

maleic anhydride, 58, 66
mean free path, 104
methacrylic acid (MAA), 68, 114, 115,
        125, 126, 145
microhematuria, 59
Mikel's reaction, 145
monofilament, 27
multifilament, 64

nasogastric tube, 48

ophthalmic polymer solution, 33
organosilicone, 27
osmolality, 49
ozone, 114, 115, 128

PE (polyethylene), 23, 45, 62, 63, 82,
        91, 92, 93, 115−121, 122,
        123, 124, 126, 127
PEG [poly(ethylene glycol)], 47, 53,
        67, 69, 78, 79
pendulum method, 28
PEO [poly(ethylene oxide)], 47, 48,
        50, 112
periodate, 128
periurethral contamination, 58
peroxide, 53, 81, 113, 114, 115, 119,
        120, 121, 123, 126, 127, 128

PET, 26, 53, 91, 114, 120, 121, 128,
  129, 131, 132, 140, 141,
  143, 145, 146, 160, 161
PHEMA, 55, 100
phosphatide, 64
photocuring, 125
photodegradation, 127
photograft (photoinduced graft), 59,
  123, 124, 125, 126, 130, 131
photoinitiator, 53, 125, 126, 130
photoirradiation (UV irradiation), 86,
  123, 124, 146, 161
photolysis (photodegradation), 127
photooxidation, 127, 132
photosensitizer, 123, 124, 126, 128
plasma flow, 120
plasma polymerization, 81
plasticizer, 59, 69
PMMA, 33, 35, 55, 65, 81, 92, 96,
  123, 128, 154
poly(dimethyl siloxane), 77, 125
poly(vinyl pyrrolidone) (PVP), 47–53,
  59, 60
polymer deposition, 81, 86
polysaccharide, 51, 66
polyurethane, 47, 48, 50, 51, 57, 58,
  60, 67, 69, 81, 113, 141,
  142, 143, 144, 147, 149,
  150, 157
povidone-iodine, 59
Prandtl-Karmer equation, 158
precorneal tear film (PTF), 33
prepolymer, 47
printability, 75, 76
protein A, 145
protein adsorption, 143, 144
protein immobilization, 88
PTFE, 13, 42, 43, 45, 76, 77, 81
PVA, 32, 33, 34, 63, 93, 113, 123
PVA hydrogel, 26, 32, 33, 34, 62
PVP-I, 59

receding contact angle, 98, 101
redox, 81
reorientation, 91, 101
Reynolds, 11, 12, 13, 154, 155, 156,
  157, 158, 159
riblet, 155
riboflavin, 60, 128, 130

saponin, 53

Schallamach wave, 6
segmental motion, 42, 73
sephadex, 142
sepharose, 88
sessile drop, 95
shear yield value, 5
silane-methacrylate copolymer, 66
silica, 112
silicone, 43, 61, 64
silicone oil, 45
silicone rubber, 23, 47, 79
silk, 23, 64, 115
SIMS, 102, 103, 105, 107
sliding angle, 26, 27, 34, 58
sodium benzoate, 123
soft segment, 47
sputtering, 105
starch, 67
stenosis, 59
Stribeck curve, 9, 153
surface tension, 34, 37, 98
surface free energy, 43
surfactant mixing, 76, 77
suture, 13, 28, 37, 38, 39, 63, 64, 65
synovia, 28, 61

talc, 67
Teslar coil, 82, 119
textured surface, 141, 143
thromboresistance, 60, 116
thrust-collar type apparatus, 147, 148
tie-down, 37, 38
titanium-nickel alloy, 57
topography, 143
trauma, 59
triboelectrification, 75, 159–161
tungsten, 58
turbulent flow, 154, 157, 158

vinyl pyrollidone (VP, NVP), 52, 56, 69

wettability, 75, 76
WLF transform, 7
work of adhesion, 43

XPS, 78, 82, 84, 85, 86, 92–95, 102,
  103, 104, 105, 132, 133,
  140, 141

yarn, 28, 39
Young's modulus, 6, 37